What Your Colleagues Are Saying . . .

If you have anything to do with transforming mathematics teaching and learning at your school, then stop everything right now and read this book. Karp, Dougherty, and Bush's *The Math Pact* strategically navigates the complex topic of creating and cultivating cohesive mathematics instruction by introducing readers to their Mathematics Whole School Agreement and specific elements that ensure school-wide success. Building on the authors' renowned *Rules That Expire* work, this book takes readers through the essential components of best practices of mathematics teaching. This should be a required read for any mathematics leadership program.

Hilary Kreisberg
Director
Center for Mathematics Achievement
Lesley University
Cambridge, Massachusetts

Just for a moment, imagine the positive and impenetrable mathematics energetic forcefield that could be created in a school if every teacher strategically and collaboratively decided upon agreements around mathematics vocabulary, notations, representations, and the overarching philosophy about how mathematics should be taught? Guess what! You do not have to imagine! No more reteaching, revising previously taught material, and explaining that "Well, in my classroom, we do it, say it, notate it this way." Because let's face it, those messages confuse students and families and can frustrate teachers. Grab this book, gather your colleagues, and get started in creating a unified and comprehensive whole school agreement that will positively enrich your students' mathematics learning experiences!

Beth Kobett
Professor
School of Education
Stevenson University
Board of Directors
National Council of the Teachers of Mathematics

As a fan of the "Rules That Expire" and "Whole School Agreement" articles, I consider this resource a welcome addition to my professional library! In this practical, easy-to-follow book, the authors provide educators with an extremely thorough and useful "go to guide" on establishing a Mathematics Whole School Agreement (MWSA). For those not familiar with the NCTM articles that initiated the movement, this resource defines what an MWSA is and why each school should establish one. More important, educators learn how to create and implement an MWSA, and the role each school and district stakeholder plays in implementation. This user-friendly resource provides teachers, teacher leaders, parents, and administrators with a comprehensive blueprint that includes research-informed practices, vignettes, grade-specific examples, and tools to encourage reflection.

Latrenda Knighten
Elementary Mathematics Instructional Coach
Baton Rouge, Louisiana

The Math Pact is a critical guide that takes stakeholders on a journey to create a Mathematics Whole School Agreement (MWSA). This journey begins with a look inward at mathematics instruction in their schools and then moves onward to compare their practices to best practices. Stakeholders arrive at their MWSA when they can ensure a unity of message that promotes coherent and effective instruction within their school and even across their district.

Juli K. Dixon
Professor, Mathematics Education
School of Teacher Education
College of Community Innovation and Education
University of Central Florida

The Math Pact is an essential resource for educators looking to develop and support a coherent schoolwide approach to the teaching and learning of mathematics. The authors provide clear guidelines with examples and references to numerous resources that will support you and your colleagues as you develop a shared vision for you and your students in mathematics education.

Mike Flynn
Director, Mathematics Leadership Programs
Mount Holyoke College
South Hadley, Massachusetts

This book brilliantly connects research-informed practices to empower stakeholders in engaging students in meaningful mathematics through a vertically articulated Mathematics Whole School Agreement! Building on the impact of the *Rules That Expire* series, the authors lay out an easy to implement approach to share, connect, and represent mathematical ideas across classrooms to intentionally and explicitly bring about change prior, during, and after instruction.

Farshid Safi
Mathematics Education
School of Teacher Education
University of Central Florida

The Math Pact will surely be hailed a seminal work in the field for years to come! The authors outline a crystal-clear, research-supported case for a unified approach to mathematics instruction. The embedded reflection opportunities and suggested parent communications make a Mathematics Whole School Agreement possible for everyone.

Delise Andrews
3–5 Mathematics Coordinator
Lincoln Public Schools
Lincoln, Nebraska

Maya Angelou said, "Do the best you can until you know better. Then, when you know better, do better." *The Math Pact* is a practical guide that supports us in collectively taking responsibility for helping each and every one of our students to become problem solvers, critical thinkers, and capable and confident doers of mathematics. The detailed vignettes and suggestions are vivid guideposts for a journey of self-reflection and collective decision making with colleagues about really critical components of mathematics instruction. It will leave you both "knowing better" and collectively "doing better" for your students. No matter your level of experience, there is something new to learn here! I think it would work beautifully with pre-service teachers, new teachers, and veteran teachers. I even picked up some new things on my read that I hadn't really thought about in my nearly 30 years in the classroom.

Shawn Towle
Mathematics Teacher
Falmouth Middle School
Past President
Association of Teachers of Mathematics in Maine

Wow! *The Math Pact* will lead the movement to help educators and students overcome the idea that math is a mysterious set of "tricks and shortcuts." The Mathematics Whole School Agreement process provides the steps, language, representations, and knowledge to build, implement, and sustain equitable learning outcomes for all students! The power to make change is in our collective hands and hearts! This book needs to be in the hands of all teachers, district leaders, and stakeholders.

Cathery Yeh
Assistant Professor
Attallah College of Educational Studies
Chapman University

This is a long-awaited publication that will help preservice teachers, educators, and administrators of all levels and curriculum coordinators abolish the use of tricks and magic in mathematics instruction. For years we have inadvertently led students down a dead-end street in their math instruction by teaching them expiring rules, tricks, and cutesy sayings that may help them perform short-term on a test but leave them conceptually damaged in the long run.

Julie Duford
Fifth-Grade Teacher
Polson Middle School
Polson, Montana

This is the perfect balance of inspiration and practical guidance! The inspiration motivates me to work harder at collaboration with peers, building common commitment. The practical guidance helps me put the ideas into action around what specific changes will improve mathematics teaching and learning.

Lynn Selking
Mathematics Consultant
Great Prairie AEA
Wapello, Iowa

The Math Pact
The Book at a Glance

Consider this book your handbook and go-to guide for ensuring equitable, coherent instruction across grades, schools, and your district. You'll find a number of features throughout the book to aid you in your journey creating a Mathematics Whole School Agreement (MWSA).

FIGURE 2.1 • WORDS THAT EXPIRE IN ELEMENTARY SCHOOL

Words that expire	Expiration details	MWSA-suggested alternatives
General		
"Show your steps"	"Show your steps" suggests that the student should be carrying out a procedure.	Instead, we recommend saying "Explain your thinking," as this phrase is inclusive of multiple options of the possible mathematical representations (e.g., concrete models, illustrations, words, graphs, symbols) and multiple strategy options.
Numbers		
Using the words take away as the generic way to read a subtraction sign in an equation—such as 14 − 8, read as "14 take away 8"	Not all subtraction problems are take-away situations and thus should not always be read that way.	Instead, simply use minus when reading such an expression or equation. Other ways to describe it include "14 subtract 8," "8 less than 14," or "the difference between 14 and 8."
Calling zero a placeholder	A placeholder is something that stands for something else. Zero is not a placeholder for another number.	Zero is a number, and as such it is a value that may in some cases represent no units or no tens, no tenths, no hundredths, no hundredths, and so on in the decimal representation of the number.
Reading a multidigit whole number such as 123 as either "one, two, three" or "one hundred and twenty-three"	Reading a number by its digits only does not promote understanding of the number's magnitude. When the word and is inserted, it implies that the number consists of a whole and a part, as in a decimal or fraction.	123 should be read as "one hundred twenty-three." The same is true for other multidigit whole numbers—no and. Meaning must be developed from the start, and there is no place value meaning given by calling out digits. However, the word and can be stated when you are reading a number that has a decimal point (as in 2.45 being read as "two and forty-five hundredths" or $9.26 as "nine dollars and twenty-six cents") or a mixed number such as $3\frac{1}{2}$, read as "three and one half." When people in the media read a multidigit whole number and say, for example, for the year 2021, "twenty, twenty-one" or "two thousand and twenty-one," we hope our students catch those and say "No and!"
	Bigger and smaller are often used when stating…	

In-depth charts will help you find a consistent approach to preferred and precise mathematical language, notation, representations, rules, and generalizations that will help clarify students' mathematics understanding.

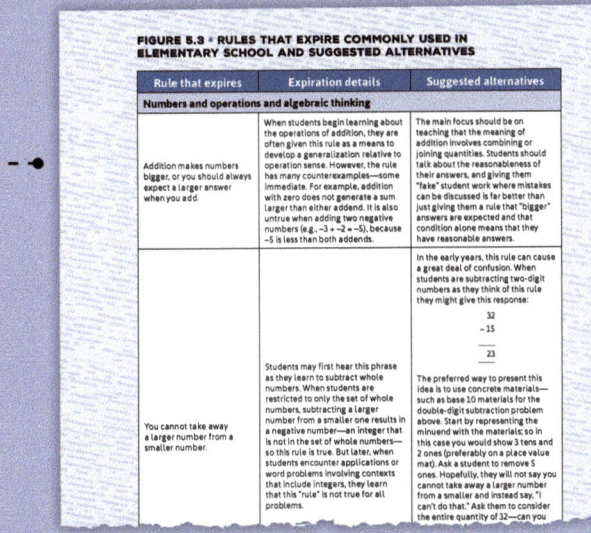

FIGURE 5.3 • RULES THAT EXPIRE COMMONLY USED IN ELEMENTARY SCHOOL AND SUGGESTED ALTERNATIVES

Remember, as you work through this chapter, you're actively establishing the RTE component of your MWSA—you're making great progress!

WHAT ARE RTEs?

RTEs are a deeply rooted tradition in mathematics education, a means to teach a procedure or a strategy in a way that the teacher believes makes the learning easy and fast or helps students remember. Sometimes RTEs are used with the best of intentions as an attempt to make learning "fun." However, let's be clear: RTEs are harmful in the long term and should not be used. We authors learned this the hard way by teaching these rules in our classrooms only to regret it later when we taught other grades or learned more mathematics content. RTEs might temporarily seem to help in the short run, but in the long run they support the myth that mathematics is a set of disconnected tricks and shortcuts, is magical, or at worst is incomprehensible.

The basic premise of RTEs is to teach for convenience or speed, and the subsequent initial appearance of student success fuels the continuance of teaching these rules. In other words, being able to apply RTEs by rote may get students through the next problem, quiz, test, or high-stakes assessment, making it seem as though there is deep conceptual understanding (or a strong reason to teach this way) when often there is not. Then, when that appearance of success leads us to believe that students understand more than they do, we use the RTEs again. In essence, the use of the "trick" or the "shortcut" becomes a self-fulfilling prophecy. Instead, we should teach for the future mathematics we know is coming and emphasize enduring understanding and

Rules that expire: Tricks, shortcuts, or rules that are used in mathematics that immediately or later fall apart or do not promote mathematical understanding.

CORE MWSA IDEA

Even actions we take as teachers that seem well-meaning can be harmful in the long run!

CORE MWSA IDEA

Throughout the book, find definitions of key terms and notes on core MWSA ideas.

REFLECTION

CONSTRUCTION ZONE–WHAT REPRESENTATIONS ARE MOST BENEFICIAL AND SPAN THE GRADES?

As you think about the representations you will use as part of your MWSA, consider these questions:

- Which representations can you agree on that will span multiple grades?
- Which representations have you used that are not productive in terms of helping students learn or for which you may not know all the options for using them?
- Which representations might cause confusion or create or perpetuate misconceptions?

Using the following space, record representations that are being used that need to be rethought, those that might need further explanation, and others that can and should be used across the grades. Then, as you continue reading this chapter, other suggestions may help you spark new ideas or prompt you to reconsider what can be used as appropriate alternatives.

Reflection tasks help you consider how key ideas relate to your own instruction.

⇨ TRY IT OUT ⇦
MWSA HANDOUT FOR REPRESENTATIONS

Representations We Are Using in _____

Representations that may cause confusion	Agreed-on representations in our whole school agreement

Try It Out and Things to Do sections provide concrete opportunities to directly engage with your team in creating a Mathematics Whole School Agreement.

THINGS TO DO

Send the Letter

Hello _____,

We have already written to you about the Mathematics Whole School Agreement (MWSA) that we are developing across the entire school this year. As you know, we are all working hard to align our instruction in mathematics across the grades. As you may remember, earlier this year you received a letter where we talked about the mathematical language and notation we use during instruction. We are now looking at the representations we use in mathematics. As a mathematics team, we have agreed on the physical materials we may use to model the mathematics and the ways in which we explain the mathematics by means of pictures or diagrams and mathematical symbols. Everyone in the school involved in the teaching and learning of mathematics is using these and is focused on teaching for students' depth of understanding and connection to mathematical ideas within and across grades. The way we model in mathematics has an effect on the way students understand mathematical ideas. We want your student to become an adult who knows mathematics and will succeed in whatever they choose to do in life. We thank you for joining us in making this shift to be consistent in how we support your student as we prepare them for their personal and professional future.

Thank you for your help,

Your student's teachers and principal and members of the school community

the
MATH
PACT

ELEMENTARY

the MATH PACT

Achieving Instructional Coherence Within and Across Grades

ELEMENTARY

Featuring Rules That Expire and Other Dos and Don'ts

KAREN S. KARP
BARBARA J. DOUGHERTY
SARAH B. BUSH

Foreword by Robert Q. Berry III and Matt Larson

A JOINT PUBLICATION

NATIONAL COUNCIL OF TEACHERS OF MATHEMATICS

For information:

Corwin
A SAGE Company
2455 Teller Road
Thousand Oaks, California 91320
(800) 233-9936
www.corwin.com

SAGE Publications Ltd.
1 Oliver's Yard
55 City Road
London, EC1Y 1SP
United Kingdom

SAGE Publications India Pvt. Ltd.
B 1/I 1 Mohan Cooperative Industrial Area
Mathura Road, New Delhi 110 044
India

SAGE Publications Asia-Pacific Pte. Ltd.
18 Cross Street #10–10/11/12
China Square Central
Singapore 048423

Publisher, Mathematics: Erin Null
Associate Content Development Editor: Jessica Vidal
Editorial Assistant: Caroline Timmings
Production Editor: Tori Mirsadjadi
Copy Editor: QuADS Prepress Pvt. Ltd.
Typesetter: Integra
Proofreader: Talia Greenberg
Indexer: Integra
Cover and Interior Designer: Gail Buschman
Marketing Manager: Margaret O'Connor

Printed in the United States of America.

Library of Congress Cataloging-in-Publication Data

Names: Karp, Karen S., author. | Dougherty, Barbara J., author. | Bush, Sarah B., author.
Title: The math pact, elementary : achieving instructional coherence within and across grades / Karen S. Karp, Barbara J. Dougherty, Sarah B. Bush ; a joint publication with NCTM.
Description: Thousands Oaks, California : Corwin, [2021] | Includes bibliographical references.
Identifiers: LCCN 2020027282 | ISBN 9781544399485 (paperback) | ISBN 9781544399508 (adobe pdf) | ISBN 9781544399522 (epub) | ISBN 9781544399515 (epub)Subjects: LCSH: Mathematics—Study and teaching (Elementary)—United States.
Classification: LCC QA13 .K38 2021 | DDC 372.70973—dc23
LC record available at https://lccn.loc.gov/2020027282

This book is printed on acid-free paper.

SUSTAINABLE FORESTRY INITIATIVE **Certified Sourcing**
www.sfiprogram.org
SFI-01028

25 26 27 28 14 13 12 11 10 9 8 7

CONTENTS

Visit the companion website at
resources.corwin.com/mathpact-elementary
for downloadable resources.

FOREWORD

As educators, we often focus on the observable learning differentials between countries, states, and school districts, and between schools within a single district. With great clarity, Karp, Dougherty, and Bush demonstrate that in focusing on these differentials we are missing what is often a more significant differential—the differences in learning outcomes that exist within schools between teachers of the same grade level or subject.

Connected to and building on the latest mathematics education literature, the authors argue that it is essential that all stakeholders within a school collaboratively agree on and commit to following a Mathematics Whole School Agreement (MWSA). By making such a commitment, each school community can approach mathematics instruction in a unified and consistent manner. The fact that the authors recommend a *whole school* agreement is significant. While many schools today are engaged in professional learning community work, those communities, when effective, may only address horizontal consistency within a grade level or subject. Effective professional learning communities within an MWSA help ensure the needed vertical consistency in addition to horizontal consistency.

Even when the same curriculum and standards are used schoolwide, the outcomes students experience in different classrooms can vary greatly due to inconsistencies in notation, language, representations, instructional strategies, assessment techniques, depth of learning, and the "rules" students learn in different classrooms. It is this lack of horizontal and vertical consistency that contributes to inequitable learning outcomes in American schools. At its heart, achieving more equitable outcomes is the goal of the MWSA, and the authors provide a process to build, implement, and sustain this necessary agreement in a school, and ultimately a district.

An MWSA provides three levels of benefits to achieve necessary consistency: teacher, student, and school levels. Teacher-level benefits include support for high-quality mathematics instruction, enhanced teacher learning, increased professional communication, reduced personal isolation, and closer alignment between curriculum and assessment. Student-level benefits focus primarily on increased student success on outcomes and depth in students' mathematical understanding, which positively influence students' mathematical identity and agency. School-level benefits include a positive influence on school climate, support for innovation, a cultural shift that emphasizes equitable opportunity and outcomes, schoolwide attention on the needs of students, flattening of the power structure, and fostering of a professional culture of intellectual inquiry.

As the authors state, "An MWSA must be grounded in a schoolwide commitment to equitable and high-quality mathematics instruction." The benefits of an MWSA address access and equity by supporting stakeholders' knowledge of the promises and challenges of the students they serve, providing a sense of collaboration for addressing potential obstacles that may limit access to high-quality mathematics teaching, and creating the space and sense of community necessary for stakeholders to ensure that the allocation of human and material resources is equitably distributed and meets the needs of both teachers and students.

In a school with an MWSA, the mathematical identity and success of each and every student become the collective responsibility of every adult involved in students' learning. We encourage you to take advantage of the authors' recommendations, collaboratively build an MWSA, make a commitment to its implementation, and make a difference in the learning outcomes of the students in your school and district.

Robert Q. Berry III, University of Virginia, Charlottesville
Past president, National Council of Teachers of Mathematics

Matt Larson, Lincoln Public Schools, Nebraska
Past president, National Council of Teachers of Mathematics

PREFACE

WHAT IS THIS BOOK ABOUT?

Imagine teaching at a school where mathematics instruction is coherent, high quality, and consistent across classrooms and grade levels. No matter which teacher a child has, they are receiving the highest-quality mathematics learning experience. All teachers in the school are working together as a team, a true team that considers the success of each and every student in the entire school as a collective responsibility. As students progress through the grades and have different teachers, they see how mathematical ideas connect, and they use familiar representations and consistent and appropriate mathematical vocabulary and notation. Teaching is done in a way that develops deep mathematical understanding, and the team knows that taking more time up front to develop concepts, connections, and procedural fluency will pay off in the long run, even saving time. Both teachers and students are excited by and feel empowered by mathematics. Welcome to *The Math Pact*, where you are about to embark on creating a Mathematics Whole School Agreement (MWSA)!

WHO IS THIS BOOK FOR?

If you are a teacher of mathematics, mathematics instructional coach, curriculum leader, principal, special education teacher, paraprofessional, parent, tutor, or preschool teacher, or anyone involved in ensuring that children are successful in mathematics, we are calling your name to join this movement toward unity of message and purposeful alignment of best practices. This book is for you and the children you teach. In short, if you care about doing what is best for students, this book *is* for you!

OUR UNIQUE AND INNOVATIVE APPROACH

Actually, our approach is really just common sense. It's all about getting everyone on the same page. We are not suggesting losing the individual style teachers have or eliminating the magic of their personality infused into mathematics instruction; we are talking about best practices and precision. These are not points of academic freedom; rather, they are ways to work toward the best interests and learning of mathematics by children in preparation for the adults they will someday be, through the implementation of research-informed best practices.

WHAT INSPIRED US

Years ago Karen started talking in her presentations at National Council of Teachers of Mathematics (NCTM) conferences and institutes about the need to avoid rules that expire (if you are not sure what these rules are, keep reading). With the encouragement of Sarah and Barbara, the three formed a team to write first an article for *Teaching Children Mathematics*, followed by other grade band articles in *Mathematics Teaching in the Middle School* and *Mathematics Teacher*. Two of these articles are in the top 10 downloaded journal articles (most recent figures) published by the NCTM: "12 Math Rules That Expire in the Middle Grades" (ranked #1) and "13 Rules That Expire!" (ranked #8). In addition, "13 Rules That Expire!" was selected as the NCTM Editorial Pick of the Year for *Teaching Children Mathematics* (2015) and was reprinted in 2019 in the compilation journal, *The Best of Teaching Children Mathematics, Mathematics Teaching in the Middle School, and Mathematics Teacher on Questions, Discourse, and Evidence*. This response was very encouraging. Many people contacted us by email to say that they resonated with the very things we had learned over the years. They wanted these ideas faster than the way we learned them, which was very slowly over the years (sorry to all our former students!)—they wanted them *now*. The next step was logically to bridge these ideas across the grades, so we wrote an article about establishing the whole school agreement, which appeared in *Teaching Children Mathematics* in 2016. After these published pieces and approximately 15 presentations at conference venues and many talks in school districts, the grade-level books seemed the next logical step to share these ideas that teachers were emailing us about, tweeting us about, discussing in their own articles and citing us, and sharing in their presentations. We thank everyone for the encouragement—it led to this series.

WHAT'S IN THE BOOK, AND HOW CAN YOU USE IT?

Consider this book your handbook and go-to guide for ensuring equitable, coherent instruction across grades, schools, and your district. This book is organized into three parts. In Chapter 1, we provide an introduction that includes describing what an MWSA is and why it is critical to the success and well-being of each and every student. In Chapters 2–6 we dive into each component of an MWSA, providing detailed vignettes and suggestions as you consider and develop your own MWSA as a team. Finally, in Chapters 7–9 we delve into the enactment of an MWSA, including incorporating it into all of your team's lessons, across your school and district, and we share success stories from those who have made this work a reality, transforming the teaching and learning of mathematics in their setting.

We hope you too feel the urgency for an agreement as a way to promote students' learning of mathematics. By providing examples and stories of our own missteps and recovery, we are hopeful we can help you navigate around our errors. Via a step-by-step walk-through of the process, we want to support you in your teamwork and your personal work. Along the way we try to point to timely resources embedded in the chapters and share why this is the right approach. We recognize that this book can never address every aspect you will need to consider as you craft your own MWSA. The purpose of this book is to provide a foundation, and then you can build off this work, using additional resources to best meet the needs of each and every student in your setting.

INSPIRATION TO JUMP-START YOUR WORK

Collaboration can save time rather than suck up time if people work productively toward common goals. The power to make change is in your hands and in your heart. It takes both hands and heart to work through the decisions that must be considered in the activities and reflections in this book that will guide your thinking. Chapter by chapter we will unveil a twofold process—one that is founded on the forging of a team and the other a self-guided and self-empowered learning opportunity. Pause and reflect, but do make change in support of having all students experience the joy, wonder, and lifetime usefulness of mathematical understanding.

ACKNOWLEDGMENTS

Our goal is to provide support for all learners by ensuring that students have a seamless and cohesive mathematical learning experience, where mathematical language, notation, representations, rules, and generalizations flow from grade to grade. Embarking on a Mathematics Whole School Agreement (MWSA) is a collaborative and exciting endeavor. This book would not have been possible if it weren't for the many partners and collaborators we've had along the way. They have been the inspiration for this work.

We wish to thank Bob Ronau, who created several of the figures for this book. A special thank you to Richard Cox, Megan Wise, Angela Torpey, Erin Russo, and the teachers and administrators of the Discovery School, Old Mill School, and Sigsbee Charter School. We also wish to thank those who contributed to this book but wish to remain anonymous. These are the stories that make our work worthwhile.

We want to thank the fans of the MWSA who work across North America, who stay in contact with us and provide heartwarming stories of how they have become revitalized by the act of the agreement process and the establishment of cohesiveness.

We are very thankful to our publisher at Corwin, Erin Null. She would come to our sessions at the conferences and share how important she felt these ideas were for mathematics education. She loved that we were asking schools to seek harmony over many significant mathematics teaching components, such as language and rules. Her own experiences as a parent and her thoughtful comments helped us look at our approach with a new lens. We are grateful for her keen insights and enduring support.

Thanks must also be extended to Jessica Vidal for orchestrating the structure of the project and for translating the family letters and activities into Spanish. Additional thanks go to the entire Corwin team and everyone involved in our book's production, which includes Caroline Timmings, Tori Mirsadjadi, Talia Greenberg, Gail Buschman, Margaret O'Connor, as well as QuADS Prepress and Integra.

We would also like to thank Robert Q. Berry III and Matt Larson, both influential past presidents of the National Council of Teachers of Mathematics, for writing the foreword for this book. Their background of national leadership roles in K–12 education with a focus on equity linked them logically to this project. We are grateful for their support and their insights into this work.

Additionally, we would like to thank all of the reviewers listed in the next section for their thoughtful reviews and extremely valuable feedback, which greatly informed and enhanced the final version of this document.

PUBLISHER'S ACKNOWLEDGMENTS

Corwin gratefully acknowledges the contributions of the following reviewers:

Susan D'Angelo
Math Curriculum Specialist
Sarasota County Schools
Nokomis, Florida

Julie Duford
Fifth-Grade Teacher
Polson Middle School
Polson, Montana

Sarah Gat
Instructional Coach/Teacher
Upper Grand District School Board
Ontario, Canada

Christine Koerner
Director of Secondary Mathematics and K–12 Computer Science
Oklahoma State Department of Education
Norman, Oklahoma

Cathy Martin
Executive Director, Curriculum & Instruction
Denver Public Schools
Denver, Colorado

Jennifer Newell
Mathematics Curriculum Development Specialist
Istation
Plano, Texas

ABOUT
THE AUTHORS

Karen S. Karp is a professor in the School of Education at Johns Hopkins University. Previously, she was a professor of mathematics education in the Department of Early and Elementary Childhood Education at the University of Louisville, where she received the President's Distinguished Teaching Award and the Distinguished Service Award for a Career of Service. She is a former member of the board of directors of the National Council of Teachers of Mathematics and a former president of the Association of Mathematics Teacher Educators. She is a member of the author panel for the *What Works Clearinghouse Practice Guide* on assisting elementary school students who have difficulty learning mathematics for the U.S. Department of Education Institute of Educational Sciences. She is the author or coauthor of approximately 20 book chapters, 50 articles, and 30 books, including *Elementary and Middle School Mathematics: Teaching Developmentally, Developing Essential Understanding of Addition and Subtraction for Teaching Mathematics*, and *Inspiring Girls to Think Mathematically*. She holds teaching certifications in elementary education, secondary mathematics, and K–12 special education.

Barbara J. Dougherty is the director of the Curriculum Research & Development Group and a professor in the College of Education at the University of Hawai'i. She is a former member of the board of directors of the National Council of Teachers of Mathematics. She serves on the author panel for the *What Works Clearinghouse Practice Guide* on assisting elementary school students who have difficulty learning mathematics for the U.S. Department of Education Institute of Educational Sciences. She is the author or coauthor of approximately 22 book chapters, 28 articles, and 35 books, including B^{Hold} *Explorations to Promote Algebraic Reasoning*. She also served as the series editor for the NCTM series *Putting Essential Understandings Into Practice* and grade-band editor for *Essential Understanding for Teaching and Learning*. Her research, funded by more than $11.5 million in grants, emphasizes supporting students who struggle in middle and high school, with a focus on algebra. She holds teaching certifications in middle and high school mathematics and K–12 special education.

Sarah B. Bush is an associate professor of K–12 STEM education and the program coordinator of the mathematics education track of the PhD in education at the University of Central Florida in Orlando. She received her doctorate in curriculum and instruction with a specialization in mathematics education from the University of Louisville. Since 2010 Dr. Bush's productivity includes more than $4.8 million in externally funded projects, nine peer-reviewed published books, more than 70 peer-reviewed publications, and more than 100 peer-reviewed and invited international, national, regional, and state presentations. She is actively involved in the National Council of Teachers of Mathematics (NCTM), currently serving as an elected member of the board of directors (2019–2022). Dr. Bush was the lead writer of NCTM's *Catalyzing Change in Middle School Mathematics: Initiating Critical Conversations*, published in 2020. Her scholarship and research focus on deepening student and teacher understanding of mathematics through transdisciplinary STE(A)M problem-based inquiry and mathematics, science, and STE(A)M education professional development effectiveness. She holds a teaching certification in 5–12 mathematics.

JUMPING ON BOARD

What Is the Mathematics Whole School Agreement?

Have you ever walked through classrooms in your school and looked at the items on the wall related to mathematics? Give it a try sometime, and consider what is similar and what is different across classrooms. What do you notice and wonder about? Perhaps you'll see a "Steps to Problem Solving" poster in your neighboring fourth-grade class and notice that they are using different steps from those in the poster in your classroom. Or maybe you'll see that two different first-grade classrooms have displays of possible mathematics thinking strategies on the wall but they don't match. You may see math word walls with completely different names for mathematical properties or algorithms. What, you wonder, will happen when those children move into second grade next year but their prior mathematical knowledge is substantially different? What confusion will ensue? How will the next year's teacher cope? What if that teacher is you? Or what if your job is to coach and support that teacher?

This book is designed to keep you, your colleagues, and your students away from this unfortunate, but all too common, situation.

In this chapter you will learn

- What a Mathematics Whole School Agreement is
- Why students need a cohesive mathematics instructional experience
- How equitable and high-quality instruction is at the foundation of the process

WHAT IS THE MATHEMATICS WHOLE SCHOOL AGREEMENT?

Mathematics Whole School Agreement: A unified and consistent approach to mathematical language, notation, representations, rules, and generalizations.

In this book we argue for the idea of building a Mathematics Whole School Agreement (MWSA). This initiative refers to a unified and consistent approach to preferred and precise mathematical language, notation, representations, rules, and generalizations that will help clarify rather than muddy children's mathematics understanding and increase their chances of mathematical success as they move into middle grades, high school, and beyond. In this book, we describe the need for an MWSA; we discuss what the agreement entails, including some very concrete mathematical don'ts and dos; and we share ideas about how to go about establishing and building the coordination and buy-in needed from educators and stakeholders to enact, implement, and get the best results from the MWSA.

So why the MWSA, and why now? *Catalyzing Change in Early Childhood and Elementary Mathematics: Initiating Critical Conversations* (National Council of Teachers of Mathematics [NCTM], 2020) describes the need to broaden the purposes of learning mathematics and articulates three key purposes for learning mathematics in the early years:

- Develop deep mathematical understanding as confident and capable learners
- Understand and critique the world through mathematics
- Experience the wonder, joy, and beauty of mathematics (p. 11)

These three purposes of learning mathematics embody the essence of the mathematical learning experiences we most want for our students—all of our students. They empower students as mathematical thinkers and doers, and they prepare students with the mathematical literacy needed for their professional and personal lives (NCTM, 2020). An MWSA builds the instructional foundation needed for these key purposes of learning mathematics to be realized in a way that is consistent, coherent, systemic, and systematic within grades, across the school, and, more broadly, within a district, state, or province. Establishing an MWSA ensures that each and every student has access to mathematically sound, consistent, high-quality learning experiences. What might happen if we don't establish an MWSA? Let's peek into a classroom:

A third-grade teacher, Ms. Jackson, is engaging her students in several problems about multiplication situations using an equal-sized group model. The first problem asks the students to think about how many children could fit in the school library if there were four tables with three children at each table. The second problem asks the students to think about how many children could fit in the same library if there were three tables with four children at each table.

Ms. Jackson's students are familiar with this problem type and select manipulatives from a central basket on the table, with most children choosing two-colored counters to represent the children and paper plates to represent the library tables. During small-group work time, this conversation occurs:

Robin: I'm looking at these equations we wrote (4 × 3 = 12 and 3 × 4 = 12), and I think this is the flip-flop property.

Ms. Jackson: I'm not sure of the flip-flop property; can you tell me more?

Robin: It's what we had in first grade. You can flip the numbers.

Jorge: Oh I know that one; my teacher last year called that the commuter property! She said commuters go back and forth to work and the numbers go back and forth.

Ms. Jackson: Robin, can you show me how the flip-flop property works with an example?

Robin: Sure, 2 + 3 = 5 and 3 + 2 = 5. You can flip them or flop them, and you get the same answer.

Ms. Jackson: Jorge, what about you? Can you give me an example?

Jorge: Yes. It's like Robin said. If you have four children at three tables and switch them back and forth, like a commuter, you can also put three children at four tables. See? [*He rearranges his counters once and then back again.*] I know this because my dad commutes, so he goes back and forth.

Ms. Jackson: I see. I think you are both talking about the commutative property. You are right in thinking there is something similar happening here with multiplication as there was with addition. Let's put "Commutative Property" on our math word wall with the equations you pointed out, Robin, so we can all use it when we notice this property appearing at other times during mathematics class.

This may seem like an extreme or even trivial example, but the dialogue from Ms. Jackson's class happens when students have at some point engaged in mathematics instruction where they are taught in ways that are inconsistent with what other teachers are using, are not well matched to the curriculum or standards, do not represent appropriate mathematical terminology, or suggest rules that later expire or fall apart. Have you ever seen this? The problem is that when consistent and appropriate mathematical language is not intentionally used, there is no evidence of vertical coherence.

Vertical coherence: The act of ensuring that interrelated mathematics concepts are aligned across grades.

Horizontal coherence: Being mindful of the relationship among mathematics concepts at the same grade.

That is, in successive grade levels, teachers and other students do not have a shared vocabulary or a shared understanding of how and when imprecise words are used. Even when the same curriculum and standards are used schoolwide, without intentional planning about what will be taught and how, the outcomes can be disjointed and students can become confused. Students begin to feel as though they're constantly learning something new and different. The irony is that while many schools work hard to enforce a unified approach to other educational matters across the school, the same is rarely true of mathematics instruction. Take classroom management, for example, where there are set guidelines for how students are expected to behave in classroom and schoolwide situations. School leaders and teachers wouldn't think to allow such inconsistency. Instead, they set out rules and norms for movement around the room or hallways between periods, when conversation is permitted, how to ask for help when you are not sure what you should do, and how to participate in discussions. These are agreed-on expectations that are consistent schoolwide. But we need to ask ourselves why there shouldn't be a similarly consistent agreement in place for teaching content. How much do discrepancies—and in some cases outright contradictions—in the way we teach mathematics and the words we use (e.g., *flip-flop property*) get in the way of having a coherent, high-quality mathematics program? How does this confuse and harm rather than help our children in their mathematical learning and achievement? How can we do better by our kids?

WHY STUDENTS NEED A COHESIVE APPROACH TO INSTRUCTION

The consistency of a message is important. We all know the feeling of having different people tell us different ways in which we need to do something and finding that hard to negotiate or navigate. Multiple communications to students with conflicting language and notation, representations, and rules and conventions in mathematics can cause mental conflict and stress for adults and children alike. This perpetuates the negative stereotypes about mathematics we hear so often: It isn't relevant to students' lives outside school; it's boring; it requires a "math brain"; it consists of a set of "disconnected ideas." To build a cohesive approach, we want to "maximize strategies that promote positive emotion" and diminish stress or threats that impede learning (Hardiman, 2011, n.p.). Research on brain-targeted teaching helps us understand how students sort the information they receive into whether those new pieces of data relate to prior experiences or knowledge

(Hardiman, 2012). Then the students build new ideas from there. If the information is in opposition to previous learning, there is a disconnect that can hinder learning or result in a backward step in retention of mathematics understanding. Squire (2004) suggests that how well we remember hinges on rehearsing and restating the ideas we learn as we set them into cohesive and connected long-term systems, constructing one layer of concepts on another. That can't happen if we don't present content in ways that help students find the familiar, identify patterns, and explicitly point out the connections between prior knowledge and new information (Skemp, 1978). Students need these linkages to deeply examine mathematics concepts and analyze situations through inductive problem-solving approaches rather than a strictly deductive model.

HOW DOES AN MWSA PROVIDE A SOLUTION?

The MWSA's design moves away from fragmented approaches and a patchwork of instructional language and notation, representations, rules and conventions, generalizations, and problem-solving approaches across multiple grades to channel an effort toward desired goals and objectives shared by all. It offers the consistency students need because it

- is an agreement shared by *all* stakeholders,
- helps students make sense of the content, and
- helps teachers ensure alignment to the standards and assessments for which they are accountable.

An MWSA Is an Agreement Among *All* Stakeholders

The MWSA is grounded in the idea that students learn mathematics more deeply and successfully when the school has a plan that *all education stakeholders who engage with students* know and follow. All of these stakeholders need to be aware of and ready to implement what educators in the school or district agree on the specific language and notation, representations, rules and conventions, generalizations, and overall problem-solving approaches that every educator in the building or district will use (Karp et al., 2016). This process of reaching an MWSA purposely brings together a broadly defined team of stakeholders that not only includes teachers, instructional coaches, paraprofessionals, and administrators but also involves substitute teachers, volunteers such as grandparents and other local community members, student teachers, staff, all family members, and others involved in students' learning of mathematics. By following an MWSA approach, the focus shifts to communicating as a unified whole about the discipline of mathematics and how it is best learned using research-informed practices. Without a clear agreement that is shared by the community as a whole, the result will be that every year the teaching of mathematics becomes harder and harder as students progress up the grades through different teachers and learning becomes more difficult for all students. Let's end this.

An MWSA Helps Students Make Sense of the Content

Some administrators and instructional leaders may say, "But we all have the same curriculum—doesn't that count?" And we respond, "That's a great start." (Later in this chapter we talk about schools that do not have a shared curriculum.) When teachers teach the same mathematics content and practices but use completely different instructional resources, the quality of mathematics instruction students receive will likely vary greatly and there is a strong risk of mathematics not being taught in coherent or consistent ways. This can occur both when teachers have a common curriculum but implement it very differently and when teachers do not have a common curriculum. These disjointed approaches lead to situations such as these:

- Teachers in subsequent grades believing that their students have prior mathematical knowledge that they do not possess
- Students harboring notions of disconnected mathematical relationships with gaps in conceptual continuity
- A general absence of the sense-making we'd like to develop in mathematics, which causes children to become confused and potentially dislike mathematics
- Students developing the feeling that they are not good at mathematics because what they were taught no longer holds true

No well-informed democratic society can afford that! Curricular coherence is about developing a consistent learning pathway in the school; it isn't about teachers teaching just what they know or sharing a collection of favorite activities.

An MWSA Helps Teachers Align Their Teaching With Standards and Assessments

Curriculum is different from, but informed by, the standards adopted in your setting. Although some states have officially adopted the Common Core State Standards in Mathematics (National Governors Association [NGA] Center for Best Practices & Council of Chief State School Officers [CCSSO], 2010), other states use what Opfer et al. (2016) refer to as Standards Adapted From the Common Core, and some others may use different state, provincial, district, or school standards. Regardless of the standards used, there remains much more to consider in an MWSA. In fact, there is little evidence of how standards are connected to what teachers actually do in their classrooms (Opfer et al., 2016). Standards documents themselves state that "standards establish what students need to learn, but do not dictate how teachers should teach. Instead, schools and teachers decide how best to help students acquire the content represented in the standards" (Common Core State Standards Initiative, 2016, n.p.). They go on to say, "Standards are not curricula and do not mandate the use of any particular curricula" (Common Core State Standards Initiative, 2016, n.p.). These statements are helpful because they not only honor teachers' critical function in decision-making but also expose the potential for using instructional approaches that lead to a disjointed collection of lessons. While teachers should feel empowered in determining their mathematics instruction, the effort should be a collaborative one with an emphasis on consistency and alignment. The MWSA requires that you work with your team of schoolwide stakeholders to establish a collective practice and focus on teaching in such a way that standards are implemented with depth and coherence, and the content and associated instructional practices across grades are aligned with attention to vertical or horizontal coherence.

> **CORE MWSA IDEA**
>
> While teachers should feel empowered in determining their mathematics instruction, the effort should be a collaborative one with an emphasis on consistency and alignment.

In asking teachers to know the mathematics content deeply and to effectively offer instruction to each and every student, it's important to acknowledge that many teachers are likely being asked to teach topics in ways they may never have experienced as a learner—either when they were in school or through their teacher preparation program. The difference is often more pronounced when we look at the mathematical practices (NGA Center for Best Practices & CCSSO, 2010) or the mathematical processes (NCTM, 2000), or other similar practices or processes adopted by your school, state, or province, because many teachers never experienced these sorts of standards when they were students. This challenge is compounded as some teachers are continuing to use more traditional instructional approaches to teach the rigorous ideas and concepts found in the required standards (Santelises & Dabrowski, 2015), which means that the standards may not be implemented as intended.

We (the authors) know what it is like to seek out curricular materials from near and far to help meet individual students' needs and to supplement content areas that need more attention. But searching for resources in the past often came with the luxury of sources that were well aligned with strong mathematical foundations and tended to be pointed to us via conference presentations, by colleagues who were master teachers of mathematics, or in NCTM journals where these resources and lessons were reviewed. They were often vetted through planning, analysis, implementation, reflection, and revision. In many cases, the experts were well versed in mathematics education and seen as more knowledgeable "others" who had based these resources on research or best practices. The resources were in some, but not all, cases reliably tested in classrooms, with solid results. Now the landscape is different and often involves nonvetted materials that don't always align with research, best practices, or standards. Additionally, the plethora of choices currently available feels like everyone is calling, "Look at what I think, or buy me." Kreisberg (2019) calls this freewheeling situation an "abundance of resources" (p. 1). She points to the enormous array immediately available at the click of a search term. But researchers (Iyengar & Lepper, 2000) suggest that sometimes, when what first appears to be alluring options becomes overwhelming, our decision-making can become seriously affected. This high number of choices can be debilitating when we become "too swamped to make meaning of them" (Kreisberg, 2018, p. 3).

Others are interested in exploring the effects of this smorgasbord of choices of instructional resources—such as researchers. The RAND Corporation has a standing interest in hearing from teachers in their well-known American Teacher Panel, a large group from across the United States whom they consult on a variety of issues. In one of their studies of 2,873 teachers, Opfer and colleagues (2016) found that 99% of elementary teachers said that they use materials "I developed and/or selected myself," and 96% of elementary teachers also reported that they use "materials developed and/or selected by my district." When asked about the use of resources found online, specifically the online resources they consulted most often, elementary teachers reported using, in order of frequency, google.com, Teachers Pay Teachers, Pinterest, their state's Department of Education website, and Khan Academy (see Figure 1.1).

FIGURE 1.1 • MOST POPULAR ONLINE RESOURCES REPORTED BY ELEMENTARY TEACHERS IN THE RAND STUDY

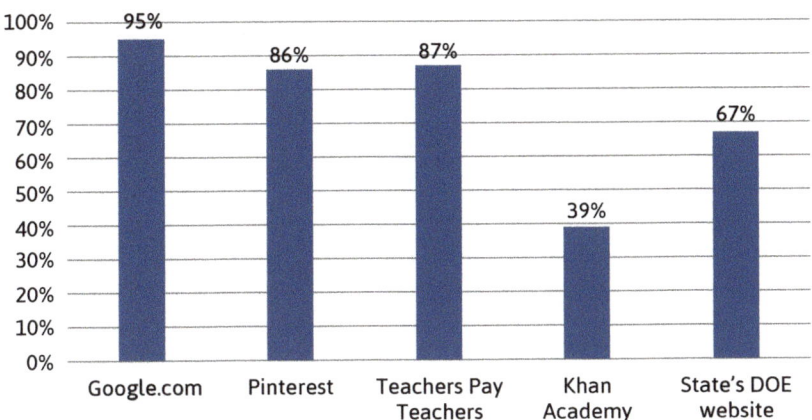

Source: Opfer et al. (2016, p. 39).

Note: DOE, Department of Education.

Interestingly, 57% of elementary teachers were required to use specific instructional materials, 27% said that materials at their school were recommended, and 15% reported having neither required nor recommended instructional materials in mathematics. It is clear from these data that teachers' use of self-selected or self-developed instructional materials is common. Furthermore, teachers reported that the factors that influenced their choices in mathematics instructional materials "a great deal" were district curriculum frameworks, maps, or guidelines; availability of materials; state mathematics standards; preparation of students for the next grade; and district mathematics assessment (see Figure 1.2; Opfer et al., 2016). Not surprisingly, they focused most frequently on the curriculum selected by the district and state standards.

FIGURE 1.2 • FACTORS THAT INFLUENCE TEACHERS' CHOICES IN SELECTING MATHEMATICS MATERIALS

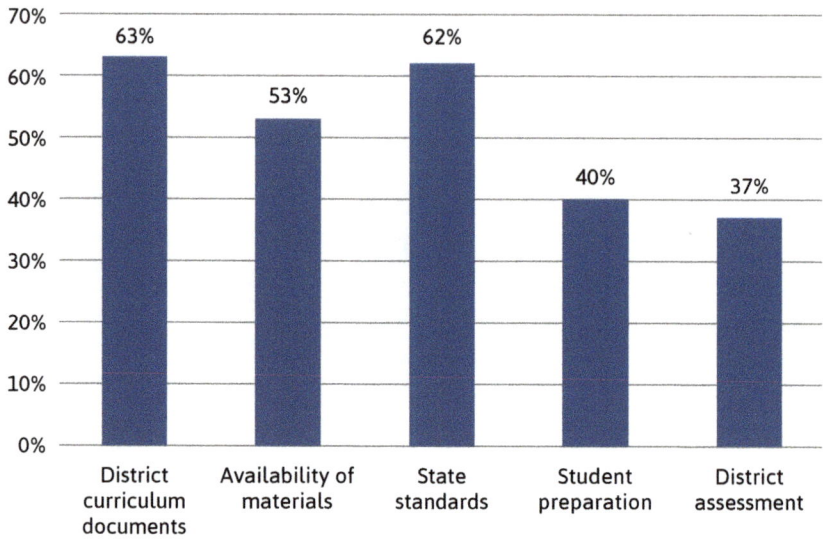

Source: Opfer et al. (2016, p. 45).

When asked if their materials provide opportunities to engage in the use of mathematical language and symbols appropriately when communicating about mathematics, 56% of elementary teachers said "to a great extent" and 49% said that they teach major mathematics topics addressed by the state mathematics standards for their grade level coherently "to a great extent." In a nutshell, this also unfortunately means that 44% of teachers did *not* report using materials that use symbols and language appropriately and more than half of the teachers did *not* agree that they teach grade-level major mathematics topics addressed by state standards in a coherent way "to a great extent" (Opfer et al., 2016). We think you'll agree that this part of the findings isn't good news.

While many schools allow and encourage teachers to self-create or self-curate the curriculum by selecting from a variety of sources, this can result in some schools having different materials used in every classroom, even within the same grade, which isn't optimal. This practice is also not an equitable, coherent, or advisable approach. Please note that we are not talking about the need to address the different learning needs of specific students. We are talking about the core curriculum. Self-curated curriculum can inappropriately create qualitatively different learning experiences for students (as described in NCTM, 2020) and is not a good use of teachers' precious time. It also runs counter to the needed approach of teachers working as a collaborative team, which fosters their professional growth and collectively benefits students. A principal who was leading a middle school in such a situation described it as follows:

> The teachers know their kids well and what the students need to know. But if I look across the mathematics program, it is "hippity skippity." By "hippity skippity" I mean that teachers who don't follow a formal program can tend to be all over the place in their pacing calendar or choice of learning materials. They rely on their own understanding of what to teach and how to teach it, which may not reflect best practices or be grounded in a recommended, research-based learning sequence.

This principal made it a point to verify that all of his teachers are trying their best, but he acknowledged that some individual teachers' decisions about selecting materials had the potential to not align with the direction of the collective group and could be out of kilter with the vertical learning articulation across grades. Selecting materials in a piecemeal way can be chaotic and cause more effort to be put into a freelance approach, with everyone rowing in different directions, than the energy required of an MWSA, where everyone is rowing on a mathematics stream in unison. When many schools are relying on a curriculum in which components are selected or substituted with different replacements by different teachers, there needs to be a decided focus on what is nonnegotiable.

What does your school agree to say and do in the mathematics classroom? This resolution can be laid out in a nonnegotiable, strong, and unified way. For example, even if something that you have decided to avoid appears in a curriculum material, you remain resolute—you collectively won't say it and won't teach it (e.g., *reducing* fractions or a keyword strategy for solving word problems). Let's map out the route to reaching such an agreement.

- To what extent are children in your setting receiving the same qualitative mathematics learning experience? For example, to what extent are third-grade students being taught the same mathematical ideas in ways that are coherent and research informed? How are the third-grade teachers coordinating with second- and fourth-grade teachers? Or kindergarten and fifth-grade teachers?
- What are some mathematics instructional absolutes that teachers in your school (or district) must follow in unfaltering ways? What practices (e.g., lecture only, teaching as telling) should be avoided?
- What are some ways to build a cohesive team of stakeholders?

COMMITTING TO EQUITABLE AND HIGH-QUALITY MATHEMATICS INSTRUCTION

An MWSA must be grounded in a schoolwide commitment to equitable and high-quality mathematics instruction. In other words, if attempting to implement an MWSA in a setting where mathematics is taught in a procedure-driven, show-and-tell, lecture format, where there is only one way to get the one right answer, this is neither equitable nor high-quality instruction. A key part and benefit for all educators of the MWSA process is learning more deeply the *what* and *how* of engaging in equitable and high-quality mathematics instruction and embracing a shared commitment to aim for this ideal. Let's break down each element a bit.

Equitable Instruction

Equitable instruction: Classroom practices that ensure that each and every student has equitable access to challenging mathematics learning opportunities.

Equitable instruction includes a commitment to developing students' positive mathematical identities and strengthening their sense of mathematical agency. This means that each and every student is seen as mathematically competent and capable and they are empowered as mathematical thinkers and doers (NCTM, 2020). Aguirre et al. (2013) define a student's mathematical identity as the "dispositions and deeply held beliefs that students develop about their ability to participate and perform effectively in mathematical contexts and to use mathematics in powerful ways" (p. 14). In a classroom in which mathematical authority is shared, students are allowed time to form their ideas and think mathematically; they engage in meaningful discourse, and their contributions are valued (Berry, 2019). Equitable instruction in the elementary grades also attends to the unique needs of young learners while aggressively working to dismantle deficit views and adopt a strengths-based approach (as described in Kobett & Karp, 2020). As simply stated in *Catalyzing Change in Early Childhood and Elementary Mathematics: Initiating Critical Conversations* (NCTM, 2020),

> We must openly challenge deficit labels and the institutional tools and practices that perpetuate static views about children's mathematical abilities and about who is or is not ready to learn. Each and every child is always ready and eager to learn more about their mathematical world. It is the adults that must reexamine their beliefs about readiness and learn to notice and support children's ever-evolving mathematical strengths. (p. 32)

Instructional practices can have both equitable and inequitable outcomes. Inequitable instructional practices will continue to privilege some students while marginalizing others. Establishing an MWSA is part of the hard work that must be done to make things equitable and just. An MWSA ensures that each and every student has foundational access to all of the mathematics opportunities they rightfully deserve.

High-Quality Mathematics Instruction

Planning for high-quality mathematics instruction should be wisely guided by NCTM's (2014a) eight mathematics teaching practices as first described in *Principles to Actions: Ensuring Mathematical Success for All* (see Figure 1.3). The eight mathematics teaching

FIGURE 1.3 • NATIONAL COUNCIL OF TEACHERS OF MATHEMATICS TEACHING PRACTICES

Mathematics teaching practices
Establish mathematics goals to focus learning
Implement tasks that promote reasoning and problem-solving
Use and connect mathematical representations
Facilitate meaningful mathematical discourse
Pose purposeful questions
Build procedural fluency from conceptual understanding
Support productive struggle in learning mathematics
Elicit and use evidence of student thinking

Source: NCTM (2014a). Reprinted with permission from *Principles to actions: Ensuring mathematical success for all*, copyright 2014, by the National Council of Teachers of Mathematics. All rights reserved.

practices inherently represent effective, high-quality, student-centered instruction and should be at the foundation of any mathematics program establishing an MWSA. When these practices are implemented systemically, systematically, and equitably across a school, each and every student can have access to a high-quality mathematics program. To guide professional learning of the eight teaching practices in your school, *Taking Action: Implementing Effective Mathematics Teaching Practices in K–Grade 5* (Huinker & Bill, 2017) provides an in-depth discussion and examples from classrooms for each of the eight teaching practices.

PRIORITIZING THE DEVELOPMENT OF DEEP MATHEMATICAL UNDERSTANDING

An essential foundation for any MWSA is a commitment to developing students' deep mathematical understanding of both conceptual and procedural knowledge. Ensuring that students develop deep mathematical understanding requires a commitment to teaching in a way that builds procedural fluency from conceptual understanding (NCTM, 2014b). Students should be "doing mathematics" in ways that focus on (a) reasoning and sense-making, (b) the mathematical practices or processes adopted in your setting, and (c) grade-level college and career readiness standards. Students should be doing mathematics (as described in Smith & Stein, 1998) through the implementation of tasks that are cognitively rigorous and relevant, offer various solution approaches, and enhance students' sense-making of a variety of mathematical ideas. (For more information on developing students' deep mathematical understanding, we suggest reading Chapter 5 of *Catalyzing Change in Early Childhood and Elementary Mathematics: Initiating Critical Conversations* [NCTM, 2020].) Along the way in this book, you will likely find times when you and your team need to brush up on the content knowledge and pedagogical content knowledge (PCK) needed for teaching mathematics. We suggest exploring the grades PK–2 and 3–5 books from the two NCTM series *Developing Essential Understanding* (2010–2013; content focused) and *Putting Essential Understanding Into Practice* (2013–2019; PCK focused). An MWSA should be built around a schoolwide instructional plan that aligns with the professional commitment of all teachers of mathematics to developing students' deep mathematical understanding. Making this pledge means avoiding disjointed and surface-level changes (e.g., using consistent vocabulary but not engaging students in deep conceptual learning) that will ultimately not prepare children for their mathematical future.

CORE MWSA IDEA

To change your practice, you have to practice change!

THE MWSA PROCESS

As we move to accept the thinking that change is not a passing fad that will simply disappear but, rather, something that benefits all players permanently, we will discuss two main components of the MWSA. First, we will detail the following central components of what all teachers and other stakeholders are agreeing to (see Figure 1.4):

- Correct and consistent language (Chapter 2)
- Precise notation (Chapter 3)
- Cohesive and consistent representations (Chapter 4)
- Evaluating rules that expire (RTEs; Chapter 5)
- Building generalizations and developing instructional strategies (Chapter 6)

FIGURE 1.4 • CENTRAL COMPONENTS OF AN MWSA JOURNEY

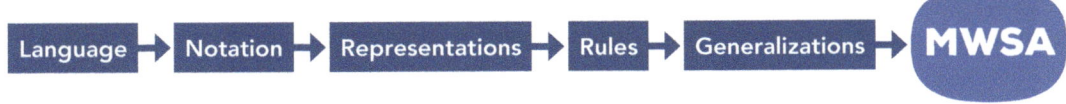

Note: MWSA = Mathematics Whole School Agreement.

Then, after you read Chapters 2–6, you'll be immersed in the second component of MWSA (Chapters 7–9), which is the last step in Figure 1.4 and an expansion of the agreement process discussed above, including everyone's commitment to it, their willingness to make change, effective instructional strategies, the structure of the lessons, and the eventual outreach to others. Not only does this process involve teamwork in structuring MWSA-aligned instruction, but you'll also explore next steps for expanding and refining this MWSA work and ensuring long-term sustainability.

In the following Reflection, predict what might be the easiest pieces for colleagues to agree to. How will exploring the next five chapters support your school as you consider developing an MWSA?

REFLECTION: **MWSA–FORECAST**

Think about the next five chapters, which will form the foundation of your MWSA. Here are some prompts to spark beginning discussions in your professional learning team or as coaches or mathematics leaders begin to think about implementing these ideas:

1. Who might you enlist as early adopters to help build your MWSA team?

2. What are some strategies you might use to gain buy-in from those who are initially resistant to the idea of an MWSA?

3. What do you think will be the easiest aspect of the MWSA for your school to agree on?

4. What are some potential challenges for both veteran teachers and novice teachers that you can predict?

5. How might the MWSA be integrated with your current curriculum materials in the school?

6. How might the MWSA lead to work that is more aligned with your content standards and mathematical practice or process standards?

7. What materials do you forecast you will need to implement the MWSA?

The following template will travel with you throughout the book. We show it here as a starting point to jot down notes as you move through the various chapters. What will you commit to in each component? Then you can partner with others and eventually discuss as a whole group what will go into your MWSA. Keep a copy of this form in your book, as it will serve as a reminder to answer the question "What will you commit to?" The more each person agrees to make changes, the stronger your agreement and your school, and your students' mathematical knowledge will be. Let's jump on board!

 TRY IT OUT

Name: _____   Grade: _____

Language

Notation

Representations

Rules

Generalizations

Instructional strategies

Lesson structure

Source: Template inspired by Karp, K., Bush, S. B., & Dougherty, B. (2016).

PUTTING IT ALL TOGETHER!

In the book *The Multiplier Effect: Tapping the Genius Inside Our Schools* (Wiseman et al., 2013), the authors describe the characteristics of people who are either multipliers or diminishers. They suggest that when people take on the role of multipliers they can build the "collective, viral intelligence in organizations" (p. 19). Multipliers will try to implement the MWSA and gather together as a force all those who are engaged in teaching children mathematics, to build over time the strengths of each and every student and child. This approach of multiplying the talent of teachers "generates the collective will and stretch needed to undertake the most paramount of challenges" as they invest in a collectively agreed-on cause (Wiseman, 2017, p. 126). In this case the cause is developing mathematically literate members of a democratic society who are well positioned to make contributions to their communities and workplaces and who feel empowered to make the world a better place.

NEXT STEPS

Now that we've started on this journey, you are seeing the full landscape of the task ahead. What stands out to you about the MWSA? What surprises you? What makes sense to you and resonates with your teaching approach? What worries you? Who is the first person you will ask to join you on this quest? Continue this journey with us as we launch into establishing your MWSA with correct and consistent mathematical language in Chapter 2. We will investigate strategies for developing a common language and notation for the elementary grades. We will also consider how these beginning steps will shape the process you will use throughout the MWSA in getting your team talking about the mathematical ideas and solidifying the ways in which decisions will be made.

WATCHING WHAT WE SAY!

Using Correct and Consistent Language

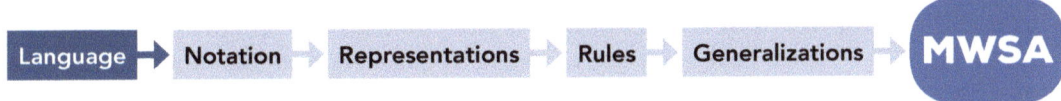

Language → Notation → Representations → Rules → Generalizations → **MWSA**

Let's revisit your own elementary school experience. For example, were you taught to *reduce* fractions? Somewhere along the way, you probably recognized that *reducing* is not what is actually happening. The fractions are not getting smaller or going on a diet; they are being written in an equivalent, simplified way. This language causes students to misunderstand the concept—especially as they link the word *reduce* with more conversational language. In this book we are calling phrases such as *reducing fractions*, *borrowing*, or *carrying* "words that expire" because they actually become a problem for students either immediately or within a few grades. We've even noticed students in high school who remain confused by language learned in elementary school. It's time to clean up our language!

Central to an effective MWSA is ensuring that across all grades and in all mathematical learning experiences students hear, see, write, and use correct and consistent language, vocabulary, terms, phrases, and labels. When the language and symbols students use are constantly shifting with each teacher or grade level, communication about one's thinking becomes scattered and muddy rather than cogent, cohesive, and connected. Teaching precise mathematical terminology and using those terms consistently shape students' ability to express their mathematical ideas. This chapter focuses on the first component of an MWSA—that is, a commitment to correct and consistent mathematical language.

In this chapter you will learn

- Why consistent mathematical language is so important
- The challenges of words holding a different meaning in conversational language and in mathematical language
- Mathematics words that are more precise than the ones we commonly use
- How to introduce vocabulary in mathematics lessons
- How to avoid mnemonics without meaning
- How to communicate these changes to others

We are excited for your team to embark on the first component of the MWSA with us, starting now!

INTRODUCTION TO MATHEMATICAL LANGUAGE

We must remind ourselves that, as with science, mathematics has a great deal of unique academic vocabulary that does not always occur in common conversations with friends and families. Furthermore, the lines between academic technical language and informal language are also blurred by mathematics vocabulary that has different meanings in everyday usage in either appearance or sound (or both), such as the following:

- acute (a cute)
- array
- base
- cardinal
- cent (scent)
- cone
- degree
- die
- difference
- digit
- even
- expression
- face
- factor
- feet
- foot
- formula
- function
- gross
- group
- hand
- hour (our)
- intersection
- key
- legend
- mean
- meter
- model
- net
- odd
- operation
- pi (pie)
- plane
- plot
- point
- power
- prime
- product
- property
- ray
- ruler
- scale
- side
- sum (some)
- table
- tree
- unknown
- volume
- yard

For young children, even the number names can cause confusion: instead of *one*, *two*, *four*, and *eight*, students might hear *won*, *to*, *for*, and *ate*. In addition to the confusion surrounding academic vocabulary because of the different meanings words have in academic settings and in informal settings, there is the possibility that words have different meanings in other academic subjects (e.g., *scale*, *proof*) and words learned in pairs may cause confusion (e.g., *area* and *perimeter*) (Rubenstein & Thompson, 2002). To be clear, even middle school students, as well as high school students and adults, can find it challenging to navigate among words used differently in informal settings than in technical mathematical language. (For a deeper discussion of this challenge and general strategies for effective mathematics vocabulary development, see Adams et al., 2005; Bay-Williams & Livers, 2009; Livers & Bay-Williams, 2014.)

The point is that if there is enough confusion already baked into mathematical academic language because of its similarity with other English words, why would we want to compound that confusion? So let's use consistent and *accurate* mathematical terminology. Where can we begin aligning? Let's first talk about mathematics vocabulary that does not help students describe mathematical ideas, relationships, and concepts, and how we can move toward vocabulary that does. We start here because changing these words is one of the easier steps in the MWSA process. As we mentioned in the beginning of this chapter, we sometimes think about these words that do not help as *words that expire*. This is a theme you'll see reoccur in this book because there are many words, rules, and procedures that may seem reasonable in a certain teaching moment or context but are in fact not the best instructional approaches. They're generally introduced with the best of intentions. But they often "expire" or become antithetical to learning mathematics at some point in the future. Sometimes this is immediately after they're introduced, because as soon as a student learns a rule or term, they run into a counterexample. For example, a young child may be asked to look for what is similar and different in a group of shapes and then later in middle school learn that *similar* has a very particular mathematical meaning in geometry. As with this situation, sometimes this shift in meaning is one or more grades later, when what students thought was a rule no longer makes sense—for example, reading all subtraction signs in an equation as "take away" when many subtraction problems are not

CORE MWSA IDEA

Start your school's MWSA here! Language is the perfect springboard to launch your MWSA!

a situation where something is actively taken away. This "surprise" can diminish students' sense of efficacy and ability.

As mentioned in Chapter 1, we know from data that some teachers conduct a Google search, visit Pinterest, or spend money at Teachers Pay Teachers (Opfer et al., 2016), but what they find is often a very mixed bag of resources in terms of quality, with often weird and incorrect vocabulary and definitions. We have found examples that describe the relationship demonstrated by 4 + 6 = 6 + 4 as the ring around the rosie property, the flip-flop property, the peanut butter and jelly property, or the commuter (travel back and forth) property. Do these phrases resonate with the desire to have students develop a meaningful and coherent collection of mathematical ideas? We suggest they do not! Although perhaps shared with the good intention of engaging students, making pathways to answer-getting memorable or accessible, or making mathematics learning fun, unforgettable, or relatable, these approaches can quickly lead to students' failure to align this important property of addition with more sophisticated but connected ideas in subsequent years. This situation occurs when, for example, their third-grade teacher has no idea what the flip-flop property is when a student tries to relate that idea to a pattern they see in $3 \times 8 = 24$ and $8 \times 3 = 24$. A dead end of confusion can repeat again when the sixth-grade teacher cannot immediately confirm that a student understands the distributive property when they suggest it is the BABY property, which is eventually described as $B(A + Y) = $ BABY! The BABY property is never a good thing to teach.

COMMONLY USED WORDS THAT EXPIRE IN ELEMENTARY SCHOOL

Expire: Something taught that is no longer accurate or meaningful over time.

In this section we share some commonly used words that are inaccurate or inconsistent, or that expire in the elementary school years. Figure 2.1 provides a list of expired words and phrases frequently used in elementary school, explains why the word expires, and provides appropriate alternatives with explanations. This list does not include every possible word that should not be used, as we simply do not have the space and more words or phrases may be locally based or invented and used by individual teachers or schools.

FIGURE 2.1 • WORDS THAT EXPIRE IN ELEMENTARY SCHOOL

Words that expire	Expiration details	MWSA-suggested alternatives
General		
"Show your steps"	"Show your steps" suggests that the student should be carrying out a procedure.	Instead, we recommend saying "Explain your thinking," as this phrase is inclusive of multiple options of the possible mathematical representations (e.g., concrete models, illustrations, words, graphs, symbols) and multiple strategy options.
Numbers		
Using the words *take away* as the generic way to read a subtraction sign in an equation—such as 14 – 8, read as "14 take away 8"	Not all subtraction problems are take-away situations and thus should not always be read that way.	Instead, simply use *minus* when reading such an expression or equation. Other ways to describe it include "14 subtract 8," "8 less than 14," or "the difference between 14 and 8."
Calling zero a *placeholder*	A placeholder is something that stands for something else. Zero is not a placeholder for another number.	Zero is a *number*, and as such it is a value that may in some cases represent no units or no tens, no tenths, no hundreds, no hundredths, and so on in the decimal representation of the number.
Reading a multidigit whole number such as 123 as either "one, two, three" or "one hundred *and* twenty-three"	Reading a number by its digits only does not promote understanding of the number's magnitude. When the word *and* is inserted, it implies that the number consists of a whole and a part, as in a decimal or fraction.	123 should be read as "one hundred twenty-three." The same is true for other multidigit whole numbers—no *and*. Meaning must be developed from the start, and there is no place value meaning given by calling out digits. However, the word *and* can be stated when you are reading a number that has a decimal point (as in 2.45 being read as "two *and* forty-five hundredths" or $9.26 as "nine dollars *and* twenty-six cents") or a mixed number such as $3\frac{1}{2}$, read as "three *and* one half." When people in the media read a multidigit whole number and say, for example, for the year 2021, "twenty, twenty-one" or "two thousand *and* twenty-one," we hope your students catch those and say "No *and!*"
Saying *smaller than* or *bigger than* or *the bigger number* to make comparisons	*Bigger* and *smaller* are often used when making comparisons, such as in the case of area or length. *Greater* and *less* are better choices because they are more closely aligned with comparisons that involve a number line.	The preferred language here is *greater than* and *less than*. If you are talking about a specific measure, you can say *longer than* or *shorter than* or *weighs more than* or *weighs less than*.

(continued . . .)

(continued . . .)

Words that expire	Expiration details	MWSA-suggested alternatives
Saying this number is *lesser* than another	The word *less* is used for a massed amount (e.g., less to eat, less paint on the brush).	Use the word *fewer* when you have a specific number of countable things (e.g., 100 words or fewer, Ginni has three fewer cookies). Introduce the word *fewer* early in a child's mathematics learning. There are word problems starting in the primary grades that include language such as "John has three fewer books in his desk than Hilde. If Hilde has eight books, how many does John have?"
Naming base 10 materials by their shape, such as *flats*, *rafts*, *rods*, *sticks*, and *cubes*	Naming the materials by their shape does not support a focus on the unit and the proportional relationships among the shapes.	It is preferred to name base 10 materials by their value, such as hundreds, tens, and ones. If you wish to continue using them for decimals, just change the name relative to what the unit is, such as calling them tenths or hundredths. If there appears to be confusion among some students due to the change in unit, you can consider shifting to paper versions to eliminate any mix-ups in the shift in values of the materials they have been using.
Using *borrowing* or *carrying* when subtracting or adding, respectively	*Borrowing* and *carrying* focus on the algorithm and not on the actual trading of equal amounts—ten ones for one ten, for example.	Use *regrouping* or *trading* to indicate the action of exchanging or trading one place value unit for another unit.
Saying the division problem didn't work out *evenly* or *exactly*	Saying an odd number can't be divided by 2 evenly or 21 can't be divided exactly by 5 is erroneous. Both numbers can be divided into equal groups, just not without a fractional component.	Instead, we suggest you say the problem didn't result in a *whole-number answer*. You can also say that a number divides into another number *without a remainder*. What is being stated with this preferred language is that the number can't be divided by a given divisor with a remainder of 0.
Saying *rounding down* (or *up*) when talking about rounding numbers	Students often think you only either round up or round down. But sometimes the number stays the same; for example, when looking at the number 30 rounded to the nearest ten, the answer remains 30.	Just use *rounding*.
Using the word *rounding* to mean *estimating*. Using the word *guess* to mean *estimate*	An estimate is not a random guess. Also, rounding is one possible strategy to produce a computational estimate, but there are many other approaches.	Use *estimate* or *estimating*. An estimate is an educated determination of either a rough calculation or the approximation of a given quantity. The word *rounding* cannot be used interchangeably with the word *estimating*.

Words that expire	Expiration details	MWSA-suggested alternatives
Reading "=" as "makes" or anything other than "equals" (e.g., "2 + 2 makes 4" for 2 + 2 = 4)	The word *makes* encourages the misconception that the equal sign is an action or an operation rather than the representation of a relationship.	The preferred way to read the equation 2 + 2 = 4 is as "2 + 2 *equals* 4" or "2 + 2 *is the same as* 4."
Suggesting that number sentences are only equations when using language such as "write the number sentence"	Number sentences are not exclusively equations.	Number sentences can be both equations and inequalities, so give students examples of each, and carefully talk about 6 > 3 as a number sentence as well 6 + 4 = 10. They need to be accurately named.
Using the words *expression* and *equation* interchangeably	The words are not synonyms and need to be used precisely.	An expression is a mathematical phrase that groups numbers, variables, and operators and stands for a single value or number, such as 15 − 8. An equation is a statement (or a mathematical sentence) that shows equality between two expressions, such as 4 + 11 = 7 + 8 or 9 = 8 + 1.
Using the words *equivalent* and *equal* interchangeably	*Equal* and *equivalent* have two different meanings.	*Equivalent* means having equal value in different forms, such as 2 × 10 = 20 and 4 × 5 = 20. These are equivalent equations, as $\frac{1}{2}$ and $\frac{4}{8}$ are equivalent fractions and three dogs and three snakes are an equivalent number of pets. *Equal* is a relationship among quantities showing the same number, value, or measure; for example, two lengths are equal if they both measure five of the same units.
Plugging a number into an expression, equation, or inequality	*Plugging* is not a mathematical term.	The preferred language is to use *substitute* instead. Here the attention then focuses on considering values for an unknown or, in later grades, a variable.

Fractions and decimals

Using the phrase *out of* to describe a fraction (e.g., "two *out of* three" to describe $\frac{2}{3}$)	The phrase *out of* frequently causes students to think a part is being subtracted from the whole amount and therefore to mistakenly select $\frac{1}{3}$ as greater than $\frac{2}{3}$ because it only has one piece taken *out of* the three pieces (Philipp et al., 2005). Or students interpret "two *out of* three" as two separate quantities rather than $\frac{2}{3}$ of a quantity.	Instead, use the fraction and the attribute (e.g., $\frac{2}{3}$ of the length of the race)
Using the phrase *reducing fractions*	When the phrase *reducing fractions* is used, it gives the incorrect impression that the fraction is getting smaller or being reduced in size (some children will say it is going on a diet!). That is not accurate.	The preferred language is *simplifying* fractions or putting fractions into the *lowest terms*.

(continued . . .)

(continued . . .)

Words that expire	Expiration details	MWSA-suggested alternatives
Speaking about a fraction having a *top number* and a *bottom number* when talking about the numerator and the denominator of a fraction, respectively	A fraction is a number. The focus should be on one number, not two separate whole numbers. This phrase mistakenly suggests that the fraction is two quantities rather than a quantity in and of itself.	Use the words *numerator* and *denominator* when discussing the different parts of a fraction (but not too early). Focus, as you did with whole numbers, with sequentially counting fractions $\left(\text{i.e., } \frac{1}{4}, \frac{2}{4}, \frac{3}{4}, \frac{4}{4}, \frac{5}{4}\right)$ so a generalization can be made such that farther along in the count represents a fraction that is greater than one earlier in the count, and vice versa (less than).
Using the word *point* to read a decimal, such as "two point nine" when shown 2.9	The appropriate occasion to say *point* is when describing how a decimal is written (or, of course, when appropriate in a geometric context). Just saying "point" also takes away from the understanding of place value (hundredths, etc.).	Instead, a preferred way to support students' understanding is to read a decimal as a fraction. In this way 2.9 is read as "two and nine tenths." Reading decimals using this approach makes the process of converting decimals into fractions unnecessary as students have been reading them in that way from the start.
Getting rid of the decimal or *move the decimal point over*	You are not *doing away with* or *getting rid of* anything or making anything disappear.	Students are creating an equivalent equation by multiplying or dividing the original decimal, terms, and so on. Keep the focus on the properties of equality or on the meaning of the operation.

Geometry and measurement

Words that expire	Expiration details	MWSA-suggested alternatives
Talking about how shapes are *similar* when students are comparing a collection of shapes	By using the word *similar* here, there can be eventual misunderstanding of the mathematical meaning of *similar*, which will be introduced in middle school relating to geometric figures.	Instead, ask "How are these shapes the same?" "How are the shapes different?" Focus on the properties or defining attributes of the shapes that are shared or are distinctive.
Using imprecise terms, such as *diamond*, *ball*, and *box*, to name geometrical shapes	Students need to learn precise terms to build their mathematical vocabulary and begin to interpret the properties that define shapes alike or different.	Instead, use precise names such as *rhombus*, *sphere*, and *cube* or *right rectangular prism*.
Using *line* for *line segment*	A line extends infinitely in both directions; a line segment is a part of a line that connects two points.	Students, even in middle school, get these two confused. If you are discussing, pointing to, or referring to a line segment, call it a line segment. If you are working with a line, call it a line, and make sure it is drawn correctly (as a line with arrows on each end and not a segment). If you have a commercially made number line, check to see that it is a line. It may be a line segment with points on each end, but you can extend it to a number line by making an extension of the line with an arrow made by a permanent marker on each end. Distinguish both of these terms from a ray.

Words that expire	Expiration details	MWSA-suggested alternatives
Naming sides, faces, *and* edges	Students need to focus on the properties of shapes, and the precise use of these terms separates the properties of two-dimensional and three-dimensional shapes.	Make sure that when you are referring to a two-dimensional shape, you highlight the sides. But in considering a three-dimensional shape, students should be discussing the faces or edges, not the sides of the shape.
Using the word *weight* when you mean *mass*	Being precise from the start helps your students be successful in mathematics and physics.	Weight is the measure of the pull or force of gravity on an object. Mass is the amount of matter in a material. On earth we use the units of mass to measure weight, which may drive some of the confusion. When you have students comparing two objects on a pan balance to see which quantity is greater, you should be using the word *mass*, not *weight*.
Using the word *fill* to describe how to measure the area of a two-dimensional figure	It is accurate to use the word *fill* to describe how to measure volume or capacity but not area.	Use the word *cover*. Area is measured by covering a space with units of area (i.e., square feet, square centimeters, etc.).
Describing a *big hand* and a *little hand* on a clock	*Big* and *little* do not support a connection to the unit. Also, many times a second hand is as large as a minute hand in length.	Instead, say hour hand and minute hand (or second hand).
Data and statistics		
Calling all graphs with bars a *bar graph*.	There are graphs with bars that are not bar graphs.	Bar graphs are used for categorical data—such as your favorite color—histograms also use bars, but they are for representing numerical data to show distributions. Bar graphs show frequencies via the height of the bars and should have the bars disconnected by a space between (the bars can actually be rearranged). Histograms are connected bars representing frequencies in the intervals of a continuous series of numbers, and the bars cannot be moved or rearranged. Bar graphs do not have to be put in order of increasing or decreasing bars!
Using the word *average* and only referring to the mean	The word *average* can apply to at least three different descriptive statistics (mean, median, or mode).	Some use the word *average* only when they are talking about the mean of a set of numbers, but to be precise, they should indicate that they are discussing the mean.

Note: The words that expire presented in this table are, in part, adapted and synthesized from ideas we have collected from educators, as well as based on our previous work in Karp et al. (2014, 2015) and Dougherty, Bush, and Karp (2017).

Let's take a look at a beginning collection of terms (Figure 2.2) that expire in your setting and that need to be jettisoned and replaced with more meaningful and interconnected words that will play an important role in students' lifelong learning of mathematics. Add more of your own suggestions on the blank lines that fit with the needs of your own school's MWSA; we know you've heard of some of these and were taught others when you were in school. As you review the list, think about what mathematical words you may have learned as a child that need replacing to rightfully emphasize meaning and help students develop a solid mathematical grounding. What terms should you use instead? How can you use your curriculum, annual assessment, standards, and research to select the best replacements?

FIGURE 2.2 • MATHEMATICS VOCABULARY MWSA SAMPLE TABLE

Words that have expired	Words we agree to use instead
Borrowing Carrying	Regrouping
Plugging a number in an expression or equation	Substituting a number in an expression or equation
Reducing fractions	Rewriting fractions in the lowest terms or simplifying fractions
Timesed—as in "I timesed 5 and 5"—or times—as in "5 times 5"	Multiplied—as in "I multiplied 5 and 5"—or groups of—as in "five groups of 5," when students are initially introduced to multiplication
Plussing	Adding
2 + 2 makes 4 for the equation 2 + 2 = 4	2 + 2 equals 4
Goes into—as in "5 goes into 8 once . . ." Using "gozinta" when describing the division process	A better approach is to say *groups of*—as in "How many groups of 5 are there in 8?" This language supports the meaning of multiplication when students realize that 5 × 4 is five groups of four as they are introduced to division. In later grades, 20 ÷ 5 would be read as "20 divided by 5."

Introducing Vocabulary

While we advocate for specific language use in Figure 2.1, we should note that preteaching vocabulary in mathematics instruction does not work to the same effect as this method does in reading and language arts. During reading instruction, teachers may initiate a lesson by preteaching words to prepare students for the language they will encounter in the text read in class that day. This approach is not as successful in mathematics instruction. In fact, Dixon (2019) calls preteaching mathematics vocabulary one of six unproductive practices in teaching mathematics. She states that if we preteach vocabulary, "we are beginning with procedures. We are saying, 'Here is the new word and here is what it means, now let's make sense of it within our lesson.' How is that leading with concepts? It isn't" (n.p.). Mathematics is sequential in nature, and if students don't know a concept, preteaching vocabulary to introduce it often does not provide the intended outcome. In other words, preloading a lesson with vocabulary and "telling" an idea robs students of the chance to develop an understanding of a concept first and then connect that concrete idea to new words. It is better to unveil the vocabulary in an experiential way in context so that the word is clearly linked to the concept at the precise point the "naming" is needed.

What works better is to have students first develop the concept through selected tasks, questioning, and meaningful discourse—using the natural language available to them—and then apply specific mathematical language to the ideas they are generating. As students use different representations and strategies, look for patterns, and rehearse different solution pathways, they are "doing" mathematics, and that is the moment to consolidate what they are doing and learning by introducing more formalized mathematical language with new vocabulary. In other words, meaningful mathematical experiences should be the primary focus, with words developed and highlighted as needed in context. Now this is not to say you should let students' loosey goosey language run rampant. On the contrary, hearing a student use an imprecise word is exactly the teachable moment you need to introduce the precise word. Then you and your students will be able to rehearse and use that correct mathematical language consistently going forward. This will help you and your students avoid the very imprecise language that expires or that we want to move away from, as outlined in Figure 2.2.

So when you hear imprecise mathematical language being used by students, what should you do at the moment? The best approach is restating the student's comment using the correct language to replace the imprecise terminology. Use every opportunity to use precise language in explaining ideas as you provide multiple opportunities for students to communicate ideas or justify answers. Thinking back to the commutative property vignette in Chapter 1 (p. 3), if your student were to refer to the commutative property as the flip-flop property, and then you ask them what they mean and they explain something representing the commutative property, that would be the ideal time to reintroduce the correct mathematical language of commutative property to that student and the class, just as Ms. Jackson did. As you find opportunities to introduce and practice precise language, this is where the use of a math word wall becomes most beneficial, so that students have an aid to look to when trying to recall the correct language. As a gentle reminder, a math word wall includes the essential vocabulary word, a visual or example, and the definition of the term. So, for example, for the word *angle* in fourth grade, the word wall would have the word *angle*, a drawing of an angle with an arched arrow showing the spread between the two rays, and the written definition.

REFLECTION   CONSTRUCTION ZONE–MATH WORD WALL

When considering what precise phrases and vocabulary to add to your word wall, ask yourself the following questions:

- What mathematical words need to be on your wall right now?
- What have students learned lately that need to be reinforced?
- What prior knowledge is important to revisit for lessons that are on the horizon?
- What language have students been confused about or misunderstood on an assessment?
- How can you combine visuals with definitions to support your learners' appropriate use and understanding of these words?

Use this template to map out your ideas.

The bottom line is that all students benefit when we shift from just teaching vocabulary to carefully planning for "when and how to emphasize correct vocabulary and formal language" (Moschkovich, 2019, p. 105). Moschkovich (2019) goes on to say that blending the teaching and learning of mathematical language with the development of mathematical practices moves learners away from low-level thinking centering on defining words and toward higher-level thinking that enhances reasoning and communication.

AVOIDING MESSY, MISALIGNED, AND SOMETIMES MEANINGLESS MNEMONICS

Let's look at another example of how using different language when teaching mathematics each year can send a confusing and sometimes inaccurate message. A study by Andrews and Kobett (2017) found that when visiting a single elementary school, confusion reigned in the use of problem-solving posters with mnemonics. Figure 2.3 showcases what they saw as they visited classrooms throughout the building.

If you look at these posters, which may be colorful and attractive (or even free), what is the overall instructional purpose that you take away from them? Often teachers use these in an attempt to be helpful—to give students an anchor and a process for solving word problems or story problems. And there isn't necessarily anything wrong with that intention. But is it effective? What happens when a different one of these posters hangs in various classrooms throughout a building? Consider Andrews and Kobett's (2017) careful analysis (see Figure 2.4) documented by traveling grade by grade throughout one elementary school.

FIGURE 2.3 • VARIOUS PROBLEM-SOLVING POSTERS FOUND IN ONE SCHOOL

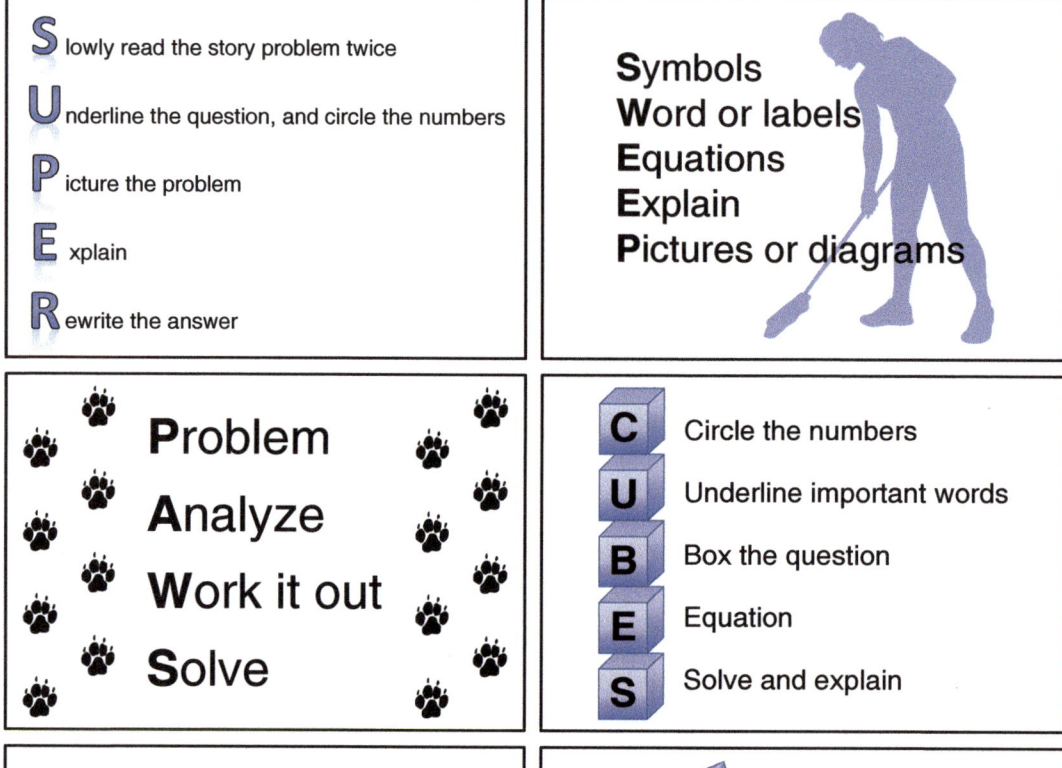

S lowly read the story problem twice

U nderline the question, and circle the numbers

P icture the problem

E xplain

R ewrite the answer

Symbols
Word or labels
Equations
Explain
Pictures or diagrams

P roblem
A nalyze
W ork it out
S olve

C Circle the numbers
U Underline important words
B Box the question
E Equation
S Solve and explain

Question
Think
Information
Plan
Solutions

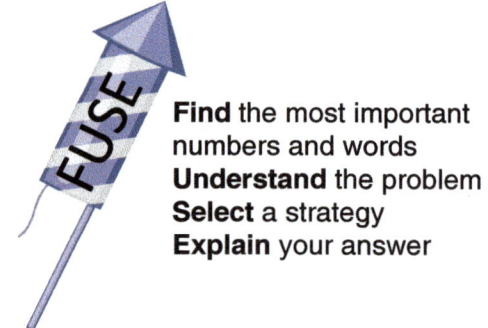

FUSE

Find the most important numbers and words
Understand the problem
Select a strategy
Explain your answer

Source: Kobett, B., & Karp, K. (2020). Image Sources: Sweeping: A-Digit/iStock.com; Cotton swab: eldadcarin/iStock.com; Rocket: Olga Andreevna Shevchenko/iStock.com.

FIGURE 2.4 • AN ANALYSIS OF PROBLEM-SOLVING PROCESS POSTERS

Grade	Problem-solving process	E means	P means	S means	U means
1	CUBES	Equation		Solve	Underline important words
2	SUPER	Explain with a number sentence	Picture	Slowly read the problem	Underline the question
3	FUSE	Explain thinking		Select a strategy	Understand
4	QTIPS		Plan	Solution	
5	SWEEP	Equation	Pictures	Symbols	

Source: Andrews and Kobett (2017).

As you might notice, the *S* on a poster in one grade means *solve*, in the next grade it means *slowly* read the problem, and in subsequent grades it can mean *strategy*, *solution*, or *symbols*. How does learning a new mnemonic in every grade help children who need support in solving problems? You're right, it doesn't. What started as an attempt to help students with a visual guide for problem-solving coupled with a mnemonic ended up being a confusing and ever-changing set of words that became arbitrary rather than infused with meaning and supportive of sense-making. That's why the whole school needs to agree that a better way to engage students' strengths would be by using a single problem-solving poster, agreed on by the whole school, that promotes sense-making. With such a consistent and agreed-on practice in place, students can reinforce what they know, find familiar patterns grade after grade, and no longer need to remember what *S* means in every grade.

CORE MWSA IDEA

If your school is going to use a poster, select a consistent and high-quality problem-solving model focused on reasoning and sense-making!

There is another mnemonic that causes problems as students begin to explore the order of operations. As students are learning about the order of operations when simplifying multistep numerical expressions, the mnemonic PEMDAS (or BEMDAS/BODMAS in Canada) is sometimes taught. PEMDAS is often expressed as *Please Excuse My Dear Aunt Sally*. PEMDAS is problematic because it encourages overgeneralizations and discourages flexibility in thinking. Instead, we suggest that students again focus on making sense of the problem rather than simply applying a rigid mnemonic. The better approach is thinking about a hierarchical model that reflects an agreement that the operations should be performed in a specified order and that we note exceptions with grouping symbols such as parentheses, as used in the United States, and brackets, as used in Canada. If we just use the mnemonic, three overgeneralizations commonly occur:

1. Students incorrectly believe that multiplication should always be performed before division, and addition before subtraction, because of how they are ordered in the mnemonic (Linchevski & Livneh, 1999).

2. Students perceive PEMDAS as rigid. For example, in the expression $25 - 8(6 + 16) + 6 \div 2$, students could actually simplify the 6 + 16 first, or distribute the 8 to the 6 and to the 16, or start with 6 ÷ 2, as any of these three options would be appropriate.

3. The *P* in PEMDAS (or the *B* in BODMAS) suggests that parentheses or brackets simplification is performed first, but the *P* should actually represent all grouping symbols, which includes parentheses, brackets, braces, square roots, absolute value, and the horizontal fraction bar.

If your MWSA team can't agree to avoid using a mnemonic for the order of operations, a better mnemonic such as GEMA or GEMS should be collectively adopted. GEMA stands for *grouping symbols* (all of them), *exponents, multiplicative operations* (i.e., multiplication and division), and *additive operations* (i.e., addition and subtraction). GEMS stands for *grouping symbols* (all of them), *exponents, multiplication or division, and subtraction or addition. These mnemonics would help alleviate the first and third overgeneralizations listed above, but a focus would still need to be placed on flexible thinking.

COMMUNICATING THESE CHANGES IN MATHEMATICAL LANGUAGE TO OTHERS

Let's take a moment to explore an important point. When we say agreed-on terms here, we mean that the whole school community agrees. In addition to teachers, that includes parents, family members, and all those who enter the building who may interact with the children. How can you best communicate these decisions to paraprofessionals, substitute teachers, student teachers, and regular volunteers? And family members? What do these important stakeholders need to know so they don't unintentionally reinforce some terms or phrases that are different from the precise mathematical language now being taught in all the classrooms in your school? An MWSA sets those boundaries for the stakeholders as a

CORE MWSA IDEA

All students need and deserve a consistent message.

cohesive whole, showing that each teacher and administrator in the school is on board. All students need and deserve a consistent message.

Before we go deeper, consider how Figure 2.2 can be used as a basis for a handout to deliver this message to other stakeholders. This component of an MWSA builds students' strengths by reinforcing the same mathematical vocabulary and ideas over and over in a unified and unwavering approach. Our goal is to further students' understanding of the concepts and the meaning of the operations and properties of mathematics. The MWSA, and the school's commitment to it, serves to facilitate that process. As you learn more about coming to an agreement on language in the rest of this chapter, think about how you will communicate the agreement to all stakeholders.

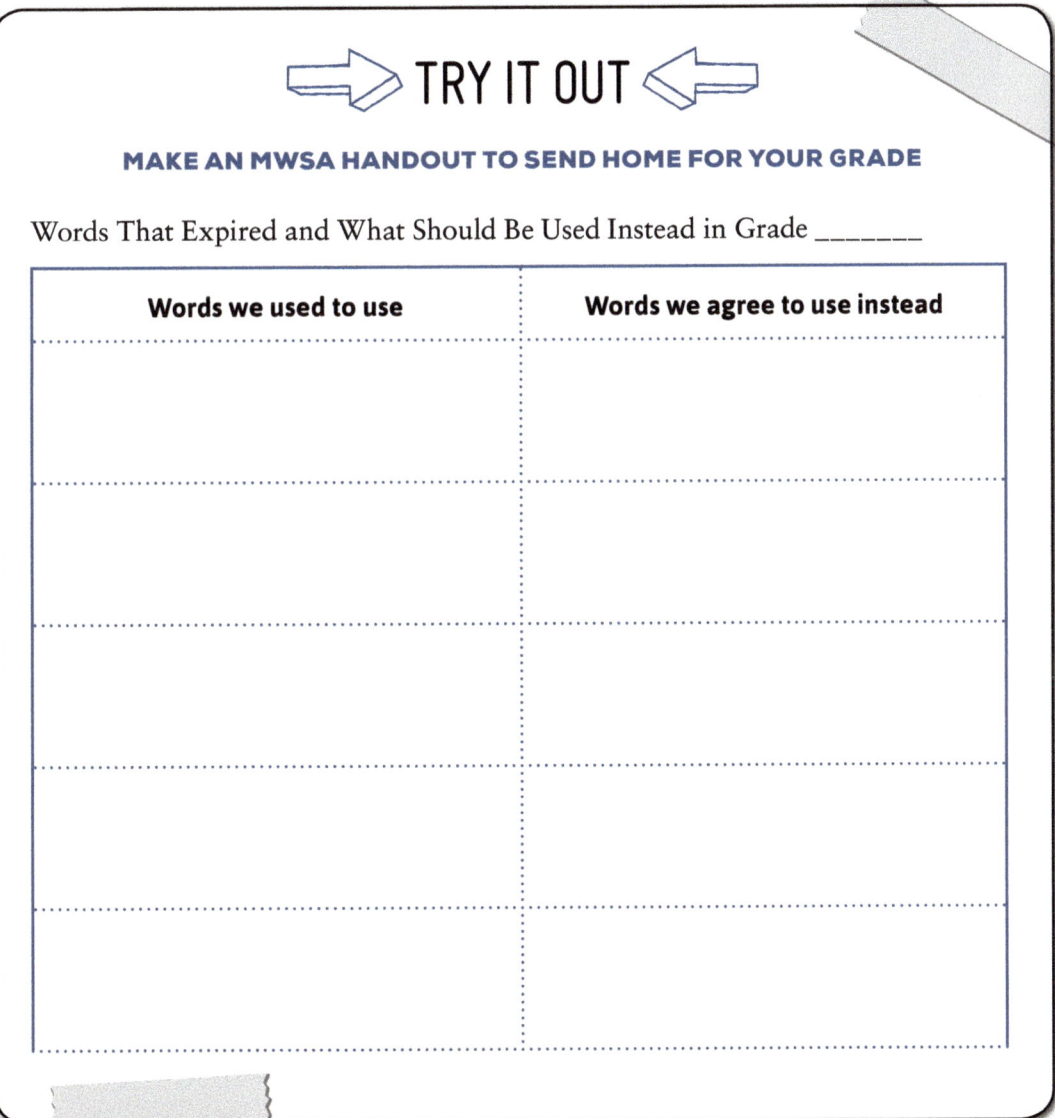

⇨ TRY IT OUT ⇦

MAKE AN MWSA HANDOUT TO SEND HOME FOR YOUR GRADE

Words That Expired and What Should Be Used Instead in Grade _____

Words we used to use	Words we agree to use instead

As you let families know about the work you are doing, here is a letter you can share to accompany your grade-level list of expired words. This message will help get on board parents, families, tutors, or others working with your students. The MWSA increases its potency with every person who consents to align with the desired (and more precise) language.

THINGS TO DO

Send the Try-It-Out Handout With This Letter

Hello Families,

We want to let you know about an important new approach we are taking at the school to support your child's mathematics learning. It is called the Mathematics Whole School Agreement. Everyone in the building is working hard to match our instructional approaches across the different grades, and we are starting by improving the mathematics vocabulary we use during instruction. We are no longer using some of the mathematical language that we, and possibly you, used when we were in school. For example, were you taught to *reduce* fractions? Actually that's not what is happening, as the fractions are not getting smaller or going on a diet; they are being simplified. So we are using the phrase *simplifying fractions*. We are working to change these often misunderstood terms in favor of more accurate mathematical language that your child needs to use now and in the years to come. Please help by using this new or revised language at home (grade-level list of expired words attached). We anticipate that you may want to ask us about other mathematics words and phrases that you think might fall into this category of outdated mathematical language. Write to us, and we will write back with ideas of stronger and more exact mathematical terminology or just say that the vocabulary you learned still works well. Expect that you may see these words in your child's assignments. Thank you for joining us in making this shift to precise and accurate mathematical language to support your learner as we prepare them for their bright futures!

Thank you for your help,

Your child's teachers and principal and members of the school community

 Available for download in English and Spanish at **resources.corwin.com/mathpact-elementary**

PUTTING IT ALL TOGETHER!

As discussed in this chapter, carefully selecting the mathematical language used in your school through an MWSA has a powerful impact on your students' understanding and readiness for the next year's content. Teachers in every grade will then know precisely what words are new and what words to reinforce, and they are assured that their students have already had exposure to a specific collection of vocabulary and mathematical phrases.

NEXT STEPS

Continue the MWSA with us as we explore *notation* in Chapter 3. We will explore how important notation is in mathematics—which is often seen as a language of symbols. While some of the notation ideas are conventions, they may be ones that we need to reinforce across all grades.

SYMBOL SENSE IS FOUNDATIONAL

Noting the Importance of Precise Notation

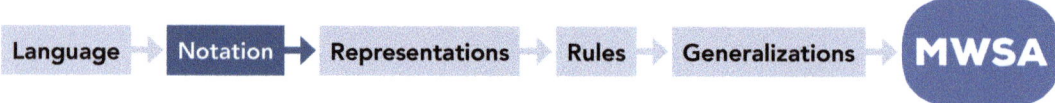

Language → **Notation** → Representations → Rules → Generalizations → **MWSA**

When many people think of the subject of mathematics, and especially if they are asked to illustrate mathematics without using words, they will likely draw numbers and other written mathematical symbols to help communicate the idea. "Symbols and symbol systems support cognitive activity by reducing the cognitive load" (Pape & Tchoshanov, 2001, p. 120) and thereby support students' development of mathematical understanding. Think how much an equal sign communicates as shorthand for a critical mathematical relationship, which will be discussed later in this chapter. Not all students can interpret the meaning of symbols without a teacher's careful development of the symbol's connection to the idea or process it represents and its precise usage or application. Some students can do mathematics when it is presented verbally but are perplexed by symbols—think of an exponent or a square root. Symbols are also a way for students to record their mental processes or actions for others to know what thinking took place. Symbols are tools for representing mathematical thoughts.

This chapter focuses on the second component of an MWSA, which is a commitment to precise and accurate notation.

In this chapter you will learn

- Why communicating symbolically is challenging
- Examples of essential symbols in the elementary grades
- Notation that we commonly use that should be taught in precise ways

As we travel through a collection of symbols in Chapters 2–4, we focus here on the second component of the MWSA—notation!

WHY THE USE OF PRECISE NOTATION CAN BE CHALLENGING

Communicating in mathematics goes beyond language, vocabulary, and phrases to the use of symbols and conventions of notation. Symbols are an essential part of mathematical understanding, but there is an incredible level of precision with symbols that can make their use challenging for learners of all ages. For instance, using the numerals 3 and 2 we can have 32, which is different from 3^2, which is different from 3.2. Each point—the position of the symbol, size of the numeral, use of italics, and so on—all go into telling a different mathematical story and point to a different representation. To complicate matters, we give symbols meaning, and although some meanings may be fixed, such as the dollar sign (\$), others such as x can have different meanings, such as a single unknown, more than one unknown, or a varying quantity. Even young learners who know the value of 5 may not be able to initially decode what the 5 in 53, $\frac{2}{5}$, 0.005, 10:5, (5,5), or 5:00 means. Reading symbols carefully and precisely is critical for the mathematical meaning to be clear; for example, 13 m^2 represents 13 square meters, not 13 meters squared. We surely want students and adults alike to know the difference between the measures represented by the symbols 0.6354 km, 635.4 m, 63,540 cm, and 635,400 mm—especially if they work in an aircraft assembly plant or with design software! To complicate the use of operation symbols, a single operation can be written in multiple ways; so, for example, 23 groups of 5 can be written as 23×5, $23(5)$, $23 * 5$, or $23 \cdot 5$, while 15 divided by 3 can be written as $15 \div 3$, $\frac{15}{3}$, or $3\overline{)15}$.

As you can see, for students to gain *symbol sense* (Arcavi, 2005; Fey, 1990), as a companion to their continually developing number sense, precision and understanding are important.

CRITICAL NOTATION TO BE LEARNED IN THE ELEMENTARY GRADES

Just as it's important for students to learn rules of grammar and punctuation, it's critical for them to also learn how to correctly communicate with mathematical symbols. We would never let children learn that a colon is the same as a semicolon or not know the appropriate use of an apostrophe, so why would we do that when teaching mathematics? Students need to grasp the meaning of symbols and carry that knowledge through their mathematics education (and lives), hence once again the need for an MWSA. Here we highlight a set of symbols that are essential to communicating the relationship between quantities.

Equal Sign

There is one symbol that permeates K–12 mathematics education that needs to be addressed immediately—the equal sign. This urgency is especially true for the youngest learners as this idea and its corresponding notation are to be mastered in most states and provinces as a grade 1 standard. Why is this conceptual understanding needed so early? Understanding the meaning of this big idea is the cornerstone that bridges arithmetic and algebra—or as it is emphasized in elementary school, number and operations and algebraic thinking. In fact, the best predictor of fourth graders' ability to solve equations is their knowledge of the meaning of the equal sign in the spring of second grade (Matthews & Fuchs, 2018). Do you assess your students on the knowledge of this essential symbol? We didn't. Many times, we as teachers just assumed that our students knew it. For many those assumptions still have not changed, but establishing an MWSA requires this nonnegotiable change.

Miller Bennett (2017) asked 42 teachers in grades 2–5 to predict the percentage of their students ($n = 1,182$) who would correctly answer the question "What is the meaning of the equal sign?" The teachers overall responded that approximately 80% of their students (differing by grade level) would be able to describe the meaning of the symbol. However, only an average of 7% of students across the grades gave an accurate answer related to the equality of the relationship between the amounts on either side of the symbol. This finding points to the need to assess your students on this important knowledge. We suggest a "well-child check" for every student in your school—just as you would take a child for any medical prevention evaluation. What is your prediction of how many students in your class, grade, or school will correctly define and apply the meaning of the equal sign? Ask them! Maybe the results will surprise you.

Researchers suggest assessing three main components of equal sign knowledge (Rittle-Johnson et al., 2011):

1. Ask students to solve an equation such as $8 + 4 = ? + 5$

2. Give students a selection of equations, particularly in nonstandard formats, and ask them if the equations are true or false

3. Ask students to define the meaning of the equal sign

Confirm that all students know the definition, and work with small groups using balances to ensure that they have a strong foundation in this fundamental notation.

Any misinterpretation of the equal sign can cause serious challenges to interpreting and reasoning about a good deal of the curriculum in mathematics studies. Students need to interpret the equal sign not as an operation such as addition or multiplication, and not as a placeholder for "the answer is" but as a relational symbol that defines the relationship of the expressions or quantities on either side of the equation. Many students still harbor a misunderstanding that is defined in the research as an operational understanding of the equal sign (Stephens et al., 2013), and they suggest that an equal sign is a signal to "do something" or that the "answer is coming up next." This notation represents one of the critical areas of student mathematical learning where the conceptual knowledge and procedural knowledge have strong interplay.

The confusion over this central idea in algebraic thinking may be driven by the consistent use of standard equation formats, such as $4 + 7 = 11$, rather than nonstandard versions such as $8 = 7 + 1$, $4 + 6 = 9 + 1$, and even $9 = 9$. Students are, for example, used to seeing an equation as representing the combining action of addition and relate it to the equation showing what is produced, rather than grasping the idea that "mathematical equivalence is the relation between two quantities that are equal and interchangeable" (Chow & Wehby, 2019 p. 678). Research by Rittle-Johnson et al. (2011) involved searching through a K–6 textbook series and found that only 4% of equations used two expressions on either side of the equal sign. In Miller Bennett's (2017) study, only 42% of the 1,182 students could accurately answer $8 + 4 = ? + 5$. Yet research reveals that solving these nonstandard equations is a foundational skill for solving the equations students will work with in algebra (Matthews & Rittle-Johnson, 2009). So in your MWSA consider the equation formats being used and when they are introduced and reinforced.

An effective way to reinforce students' learning of the meaning of the equal sign as a relational symbol—instead of the incorrect notion that it is a symbol indicating an operation—is to read the symbol accurately in all equations. So, for example, $9 + 7 = 16$ should not be read as "nine and seven make sixteen" or, even more off target, "nine plus seven is sixteen." Instead, the equation should be read as "nine and seven *equals* sixteen." Very simple! Making this shift is actually easy, as it recognizes the name of the symbol while building a consistent message. Now we must warn you that you will hear and see in print the incorrect reading of the symbol in a variety of places, including mathematics curriculum.

CORE MWSA IDEA

When reading an equation, the equal sign should be read as just "equals."

But just as with the sometimes incorrect reading of numbers on TV and in the media, we rise above this and point out the accurate way to read the equation.

Greater Than and Less Than Signs

Just as we think about the equal sign, we must also address other relational signs, such as greater than and less than. Do your students know how to distinguish signs such as $\neq$, $\leq$, $\geq$, $>$, and $<$? Additionally, can they read them properly in an inequality such as $11 > 7$ or $72 < 143$? Do they still refer to an alligator mouth as they try to read the inequalities? Eventually, students will look at inequalities such as $-12 < x < -1$ or $-4 < x + 2 < 6$ (as well as ones with the symbols $\neq$, $\leq$, $\geq$, etc.). How would they respond? Possibly by using wide-open alligator mouths? Or by holding up their left hand's index finger and thumb to make the shape of the letter L. If you are not sure what we are talking about here, it is probably best that you do not know! Again, focus on the properties of inequalities, the properties of operations, and making sense of the mathematics. The symbols we use represent the mathematical language and the ideas we want to communicate—and they are very useful as a shorter way of writing all those words.

Often teachers report that their students can figure out the symbol they should use in a given numerical comparison but they are challenged to read the inequality correctly or know what these inequalities mean conceptually. Rather than having the alligator bite the bigger piece or greater quantity, we need to teach with meaning, and if there is difficulty, we need to use a more fundamental and reliable approach to cementing this foundational pair of relational symbols. One way to define the relationship more clearly is to use centimeter cubes and a flexible angle (see Figure 3.1) that can be made larger and smaller at the vertex (Livers et al., 2019). Then actually compare numbers to link the symbol to the actual exploration of the quantities. By angling the rays of the angle to align with the heights of the two towers, the symbol appears, and it should then be written down as a record of the inequality and read aloud. If the heights are equal, two unattached angle legs can be used at the top and bottom of the towers to show the equal sign.

FIGURE 3.1 • A NEW LOOK AT THE RELATIONAL SIGNS—NO ALLIGATORS ALLOWED!

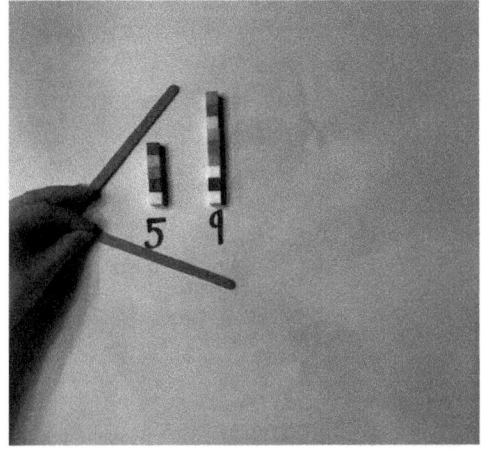
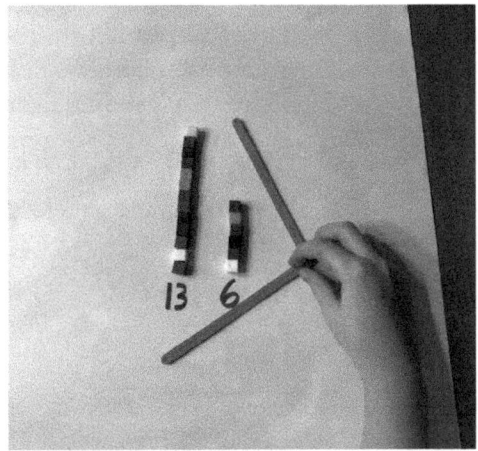
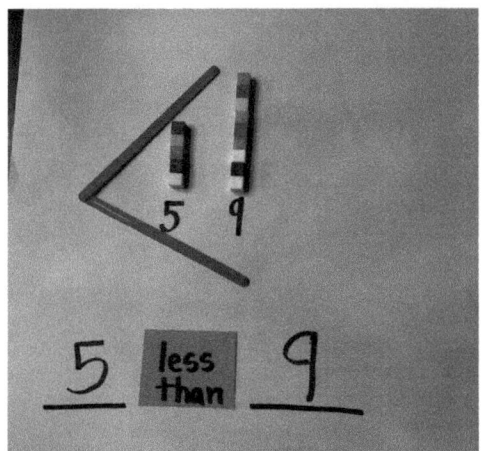
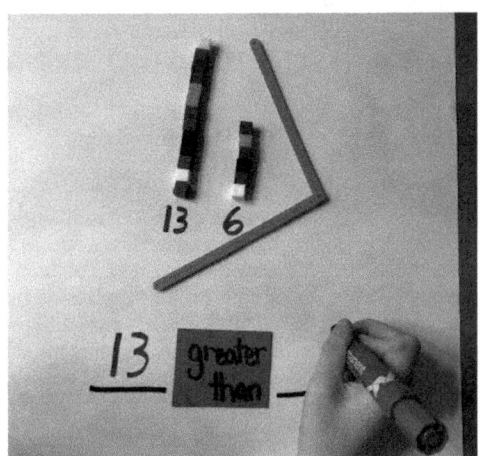

Source: Livers et al. (2019, p. 412). Reprinted with permission from Danger! Animals in the mathematics classroom. Teaching Children Mathematics, 25(7), copyright 2019, by the National Council of Teachers of Mathematics. All rights reserved.

Students might need to work toward mastering the accurate reading of the inequality, naming the values and symbols from left to right. So we don't want 13 < 25 to be read as "twenty-five is greater than thirteen," but instead we want the actual reading of "thirteen is less than twenty-five." If students incorrectly read this inequality as "thirteen is greater than twenty five," their errors will be highlighted—which is what we want, rather than having them hide behind a work-around. This, however, is a teachable moment because 13 < 25 and 25 > 13 are ways of expressing the same relationship.

COMMONLY USED NOTATION IN ELEMENTARY SCHOOL THAT NEEDS ATTENTION

Let's pinpoint other notation in elementary school that needs attention. Figure 3.2 provides a list of commonly used symbols and shares what should be used instead. These ideas need to be reinforced heavily at the elementary level so that students have the groundwork in place for success. This list is just a sampling—consider other such notation that exists in your setting.

FIGURE 3.2 • NOTATION THAT EXPIRES IN ELEMENTARY SCHOOL

Notation that expires	Expiration details	MWSA-suggested alternatives and explanations
Using the notation $18 + 4 = 22 + 6 = 28 + 13 = 41$ to symbolize a running series of addition problems	You can see the thinking here. As an answer is given, another addend is tagged on for a new problem, followed by a new answer, a new addend, and so on. Stringing together this series of additions (or other computations), the addends cannot be connected with equal signs as the computations are unequal. This string is in fact nonsensical and illogical and distorts the meaning of the equal sign as a balance.	Students must use three individual equations in this case to be accurate. For each answer from the previous problem, a new equation is created with the answer as the first addend. Equal signs must connect equal quantities.
Using a *diagonal bar* when writing a fraction	This notation is an issue when young learners read the diagonal bar as a 1 (e.g., 5/8 is read as 518, and 5/10 = 1/2, especially when handwritten, is read as 5,110 = 112). The diagonal bar is an issue later too, for example, when students see 1/3x and are unsure if it is $\frac{1}{3}x$ or $\frac{1}{3x}$. When trying to connect symbols to situations in context, precision matters.	The use of a *horizontal bar* is preferred. Instead of 1/2, write $\frac{1}{2}$ by hand on the board and in any tasks you create (e.g., Microsoft Word and Google Docs have an equation editor feature that makes this easy). The consistent use of the horizontal vinculum takes much of the confusion away. The use of the diagonal bar was mainly for typesetters, who did not want to break the line of type in print form. But it seems that they were willing to get beyond this need in secondary mathematics books, where the diagonal bar is less prevalent. Let's start early.

Notation that expires	Expiration details	MWSA-suggested alternatives and explanations
Writing the decimal seven tenths as .7	The decimal seven tenths should be written as 0.7, as writing a decimal without the leading zero does not align with international notation standards.	International notation standards indicate that a leading zero must be used when a measure can possibly be greater than 1. The leading zero is also an indication that a decimal is coming, and students are less likely to overlook the decimal point (we can all imagine that this convention is particularly critical in circumstances such as medicine doses). Many students are more successful when this notation is consistently used.
Writing only a number to represent a unit of measure. For example, when asked about the area of a rectangle that is 3 units by 3 units, a student gives the answer 9	This influences students' grasp of how to accurately use notation in mathematics. When we measure, we must use units that have the same attribute as what we want to measure. So measures of length need a tool and units that have length, and measures of an angle need a tool and units made up of angles. It is no different for measuring area.	The answer must be 9 square units. We measure area with units that cover space—commonly square units. Without the unit of measure given to notate a measurement, the answer alone is without meaning.
Using a blank space (not a blank line, an empty space) instead of using a representation for an unknown to show an equation	For clarity, there is a preference to use an unknown rather than just a blank space to separate an expression from an equation.	When writing an expression as 5 + 4, it represents an amount. Adding an equal sign indicates a part of an equation. So, for example, you could write 5 + 4 = ? instead of 5 + 4 = .
Writing a variable as "x"	Students often confuse 3x as the beginning of a multiplication problem with the multiplier missing, thinking that the x represents multiplication.	Use italics when writing variables, to help students make sense of the symbols—especially when x is used as a variable. Write 3x instead.

Source: The notations that expire presented in this table are, in part, adapted and synthesized from ideas we have collected from educators, as well as based on our previous work in Karp et al. (2014, 2015) and Dougherty, Bush, and Karp (2017).

CONSTRUCTION ZONE
REVISITED—NOTATION WALL

Now that you have a math word wall, consider these questions:

- What notation should be there for students' reference?
- What have students learned lately that needs to be reinforced?
- What prior knowledge is it important to refresh for upcoming lessons?
- As with vocabulary and phrases, how can you combine visuals with definitions to support your learners' use and understanding of these symbols?

Use this template to map out your additional ideas.

PUTTING IT ALL TOGETHER!

"It [mathematics] is both a written language and a spoken language, for—particularly in school mathematics—we have words for virtually all the symbols. Familiarity with this language is a precursor to all understanding" (Usiskin, 2015, p. 824). As discussed in this chapter, carefully selecting the mathematical notation to be used in your school through an MWSA has a powerful impact on your students' understanding of and readiness for the next year's content. There is nothing more important than to communicate mathematics clearly in the teaching and learning interaction. Precise use of symbols supports the transfer of knowledge and sharing of understanding.

NEXT STEPS

Continue the MWSA with us as we explore representations—another form of symbolic understanding—in Chapter 4. We will investigate the use of representations that can transcend multiple grades, adding more meaning and understanding to students' conceptual development.

MENTAL IMAGES THAT LAST

Cohesive and Consistent Representations

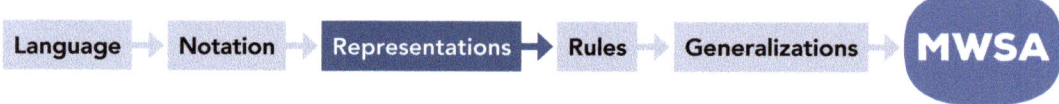

Language → Notation → **Representations** → Rules → Generalizations → **MWSA**

"I can't use concrete materials in my math class because my students will depend on them and they can't use them on the test."

"I have a lot of boxes of manipulatives in my classroom closet, but I have no idea about all of the things they can be used for."

"I didn't need any concrete materials or other representations to help me learn math, so my students probably don't need them either."

Have you heard comments such as these in staff meetings or in the teacher workroom? At the elementary school level, the use of manipulatives or any representations other than symbolic may be put aside due to "time restraints," in spite of the fact that the use of models such as connecting cubes, two-color counters, base 10 materials, and other representations in our mathematics classes strongly supports students' thinking as they develop strategies to solve problems. But aren't these materials too expensive? Are upper elementary students going to think that physical models are only for younger students? Shouldn't we be preparing them for middle school? As you read and discuss this chapter, take this opportunity to challenge the perception that elementary school students don't need multiple representations for all mathematics topics and can learn and understand mathematics by only seeing the symbols or picture in the book.

In this chapter you will learn

- Three types of representations that should be included in your mathematics instruction
- Examples of representations that may inhibit student learning

As your MWSA team decides on representations to include in your MWSA, don't hesitate to bring out manipulatives to support your discussion and to practice weaving the idea of using concrete materials into every unit. Your team is moving forward, and this step in the development of your MWSA takes you one step closer to completing it.

WHAT ARE REPRESENTATIONS?

Given the pacing of many curricular materials and the number of standards that students are expected to learn, it is understandable for us to believe that it is so much easier for students and more efficient if we just, for example, tell them how to carry out a procedure, then let them practice it to gain proficiency, and repeat as necessary. Or, as you will read in the next chapter, we may believe that providing students with a quick trick or rule to help them solve a problem will get students to the end goal of solving problems faster. There is no doubt that, on the surface, demonstrating procedures for students to mimic and offering shortcuts or tricks certainly appear to be faster ways to help students get to an answer in that moment. But is the goal of mathematics answer-getting or long-term mathematical understanding? Is the goal for students to accurately perform calculations on a test or to learn strategies or approaches that they can transfer to any situation? There are actually two dangers inherent in teaching in ways that are at the surface level:

> **CORE MWSA IDEA**
>
> **Teaching procedures and shortcuts that are disconnected from developing conceptual understanding does not lead to long-term mathematical knowledge and improved performance.**

1. When we overemphasize teaching rules and procedures—or sometimes tricks or shortcuts—we are often faced with the need to reteach again, and again, and again. When students do not attach any *meaning* to the procedures they are performing, they easily forget how to carry out a solution to a problem. Or, worse, they develop a "buggy" algorithm or solution strategy that is rife with errors yet is somehow the one they remember clearly and repeat over and over. And in the end this is not a path to achieving any of our goals faster or more efficiently; the resulting need to reteach actually causes us and our students to fall further behind. This approach undermines the very need we were trying to address: high-quality mathematics learning for every student.

2. More important, teaching students procedures and shortcuts disconnected from developing conceptual understanding does not lead to robust and long-term mathematical knowledge, enhancement of their ability to carry out the mathematical practices or processes, or improved performance. Students are often at a loss when they have to solve a problem for which

the procedure is obscure or that is changed in some slight way; they can't "see" the relationships between the concepts and the algorithm and need some way to make it more visual or connect it to prior knowledge of other content or practices.

It is beneficial to use representations as tools because they

- provide ways for students to examine and identify relationships,
- help students compare and contrast multiple depictions of a mathematical idea to determine generalizations or patterns,
- give students a way to communicate their thinking about a concept or process (Barrera & Santos, 2001), and
- support deeper understanding of the mathematical content and practices than would otherwise happen if we only present these ideas with the use of mathematical symbols while hoping that students make accurate mental connections.

REPRESENTATIONS: CONCRETE, SEMICONCRETE, AND ABSTRACT

Representations are commonly thought of in three different ways: (1) physical or concrete (e.g., hands-on), (2) semiconcrete or diagrammatic (e.g., pictorial or graphically represented), and (3) abstract or symbolic (e.g., written as symbols—numbers, variables, and operation symbols). You may be familiar with an organizational structure called the *concrete-semiconcrete-abstract* (CSA) approach (Miller et al., 1992) or perhaps the *concrete-representational-abstract*, or CRA, approach, which is the term commonly used in special education (Peltier & Vannest, 2018). Both of these approaches are simply a way of organizing these categories of representations, and we prefer to use CSA because it accurately recognizes that the three components (the *C*, the *S*, and the *A*) are all representations. Lesh et al. (1987) identified important connections among contextual, physical, symbolic, verbal, and visual representation forms. This work is embodied in the CSA approach, as physical aligns with concrete, visual aligns with semiconcrete, abstract aligns with symbolic, and verbal and contextual are interwoven throughout the CSA approach.

Concrete or physical representations are actual objects or models that students touch and manipulate. They may be objects that are found in students' everyday life, such as containers to model volume or surface area or three-dimensional shapes. Or they may be teacher-made or commercial materials, such as Cuisenaire Rods™ or two-color counters.

Semiconcrete or diagrammatic representations may be a pictorial representation of the concrete objects or a picture or

Concrete representations: Objects or models that students can manipulate to represent mathematical ideas.

Semiconcrete representations: Pictorial representations of concrete objects or a picture or graphic organizer that represents a problem situation. They also include diagrams or tables.

graphic organizer that represents a problem situation or schema. They also include other types of representations such as number lines, graphs, diagrams, and tables. Semiconcrete or diagrammatic representations are also commonly used in technology applications such as interactive number lines, dynamic geometry software, and virtual manipulatives such as tangrams or representations on interactive whiteboards.

Abstract or symbolic representations are those that incorporate mathematical symbols such as numbers, operation signs, equations, or expressions. Abstract representations are commonplace in elementary school mathematics. Furthermore, it is in middle school when many new algebraic abstract representations are first introduced, which can be a challenging shift for students without proper prior knowledge. There is also a fourth representation, already discussed in Chapter 2—the use of natural and mathematical language. Natural language can be thought of as more informal than mathematics language, and you may notice students using natural language in mathematics when they are explaining a new idea or making an observation. Mathematical language is more formal than natural language and conveys ideas using the academic vocabulary and syntax of mathematics. In a sense, two types of abstract representations, linguistic and symbolic, have already been addressed in Chapters 2 and 3, respectively.

When adopting a dedication to using the CSA approach, teachers may often start by presenting students with the physical model, then progressing to the semiconcrete stage, and finally using abstract representations (Maccini & Ruhl, 2000). This progression, however, can create a disconnect across the three types of representations, so that students are not clear on how the concrete models are related to the abstract (McNeil et al., 2009). Rather than thinking of each representational type in isolation or as separate from the others, or as a linear progression moving from concrete to semiconcrete to abstract, students understand mathematics best when all three types are presented simultaneously (Dougherty et al., 2016; Moreno et al., 2011; see Figure 4.1). The intertwining of representations leads students to create what we call a *mental residue*. That is, they create a mental image of the actions they experience with the materials, which then connects with their semiconcrete and abstract representations (Dougherty, 2008; Okazaki et al., 2006). The mental residue helps students form robust conceptions and understand the relationships that are embedded in the representations. When working with representations, using the CSA framework is a nonnegotiable to better leverage student learning!

Abstract representations: Mathematical symbols such as numbers, operation signs, equations, or expressions.

Mental residue: A mental image of physical actions done with concrete materials.

CORE MWSA IDEA

Leverage student learning by helping students create mental residue through using a csa approach.

FIGURE 4.1 • THE CSA APPROACH

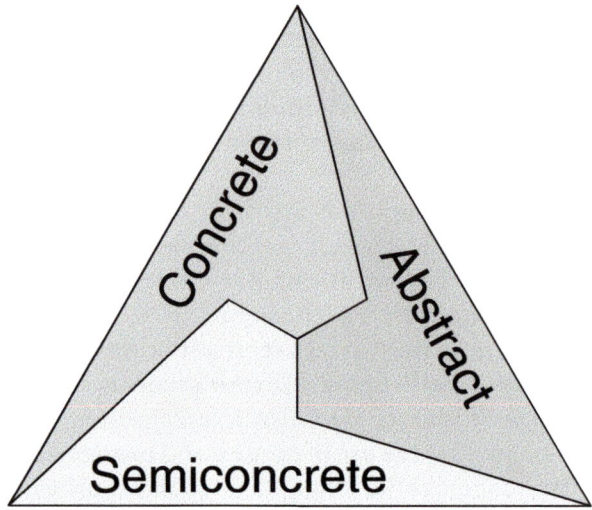

Source: Created by Robert Ronau.

SELECTING ALIGNED AND CONSISTENT REPRESENTATIONS

We've established what the CSA approach is and why you should consider using it to support students' understanding, but now the question is "How do we agree on what representations are the best ones to use?" How you choose the representations that your team will include in units is one of the critical decisions you have to make. Choosing a weak representation (or not choosing one at all!) can greatly affect student learning. And, as part of your MWSA, it is important that all the members in your MWSA team agree on which representations make the most sense to use and which ones to avoid, so that students receive cohesive instruction. So how do you know what mathematical representations to use and how to use them?

CORE MWSA IDEA

Not all representations are created equal!

Concrete Representations

In thinking about concrete and physical models, the best place to start is to bridge from what we know from research over many years. Laski et al. (2015)—who reviewed multiple studies in cognitive science— found four general principles for successfully using concrete or physical models, which apply to any grade level or mathematical topic:

1. Use concrete or physical materials consistently. The more experience students have with a particular manipulative, the better able they are to use it appropriately and make deep connections with the concept(s) it embodies.

2. Select a concrete model or representation that will clearly exemplify the abstract representation. It is helpful if the actions performed on the model or the representation itself are closely aligned with the abstract representation of, for example, the process underlying an algorithm, such as adding or subtracting multidigit numbers.

3. Avoid concrete models or representations that have distracting features that may take away from using the tool to make sense of a concept. For example, if students are working on addition problems and they are given play money just to use as counters, that might confuse them rather than focusing on the operation as an action of combining.

4. Make the relationship explicit between the concrete material and the concept being modeled. Rather than having students try to determine how the concrete material models the concept, be intentional and specifically ask questions of students that prompt them to think about how the representations are related.

Not all concrete materials accurately and appropriately represent mathematical ideas. Let's look at place value materials as an example. There are a variety of materials available to model many ideas in elementary school mathematics, such as multidigit quantities, comparisons of whole numbers, and quantities in addition, subtraction, multiplication, and division situations, many of which are helpful in supporting students' understanding. There are, however, some potential pitfalls in using maniuplatives if a good choice is not made of the representative materials or if the materials are not implemented appropriately.

CORE MWSA IDEA

Not all concrete materials enhance thinking — select and use strategically!

We illustrate a few examples of place value representations that should be avoided. In Figure 4.2, stirrers are used in a pocket chart, where the value of the stirrer changes just by changing its location. So, for example, by just moving a single stirrer to a new location, the value of the stirrer changes from a one, to a ten, to a hundred. This approach should not be used when introducing place value. Instead, bundles of 10 single stirrers should be bound together with a rubber band to be placed in the tens pocket and bundles of 100 single stirrers in the hundreds pocket. The representation needs to match the quantity in order to build conceptual understanding and help support understanding of regrouping versus the act of just carrying or borrowing a digit. Let's face it, if students could easily see that moving a single item from pocket to pocket changes its value—they are at a level of abstraction that does not require this, more concrete model!

FIGURE 4.2 • POCKET CHART WITH NONPROPORTIONAL MATERIALS

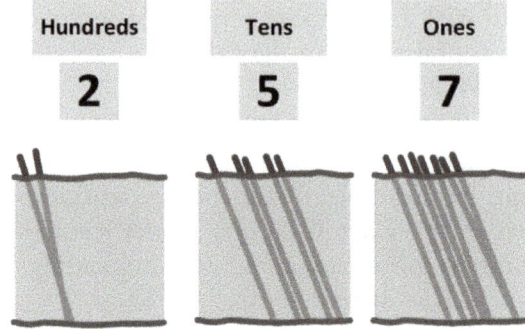

Source: Created by Robert Ronau.

In considering the next example, we suggest you save your money and not purchase this commercially offered item (see Figure 4.3). Again, this is a nonproportional model and, as such, is not valuable in building the concept of place value for beginning learners. Here, color or the number signals the chip's value, and therefore, depending on these colorings, rather than the quantity, a chip can represent a hundred, a thousand, or a one. Instead, use base 10 materials.

FIGURE 4.3 • PLACE VALUE CHIPS NONPROPORTIONAL MODEL

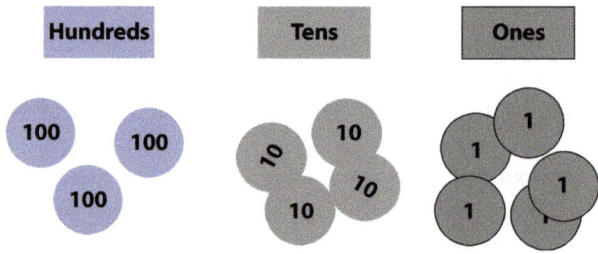

Source: Created by Robert Ronau.

Base 10 materials, whether paper, as shown in Figure 4.4, or materials such as connecting cubes, wood, foam, or plastic, are area or volume models. It will take some time to explain to students the relationship between the values. That is why it's best to start with groupable models where tens can be built—such as connecting cubes—and segue to pregrouped or trading models that are already representative of values such as ones, tens, and hundreds through the resulting area. Notice that in Figure 4.4 we show the value in symbols (200, 30, 3) of the place value materials to link the representations explicitly to the numeral 233. We also expect students to read and write the corresponding number that matches the quantity.

FIGURE 4.4 • PAPER PLACE VALUE MATERIALS SHOWING 233

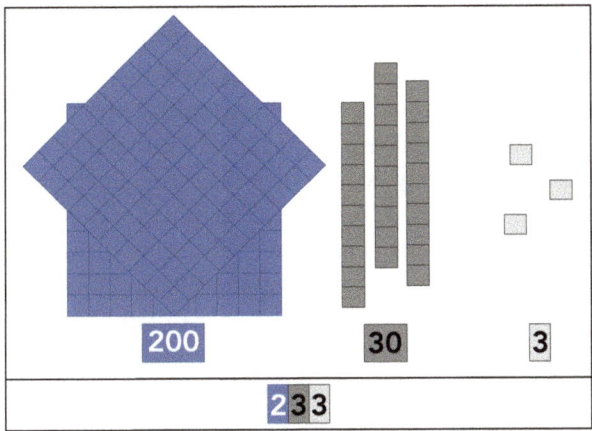

Source: Created by Robert Ronau.

Representations used strategically can counteract common errors or misconceptions. For example, using the same paper materials shown above, students can counteract a possible confusion when you ask them to represent 107. When students use the materials to represent 107 on a place value mat such as the one above and are just asked to write the corresponding numeral, they may write the number as 17, forgetting the internal zero. A better approach is to again use the number cards as shown above with the collection of options for the hundreds, tens, and ones and likewise put them together right justified to show how the number is built. When there are no tens, the 0 from the number in the hundreds shows through (see Figure 4.5). Again, the link between CSA builds a solid foundation for clarifying concepts!

FIGURE 4.5 • PAPER PLACE VALUE MATERIALS SHOWING 107

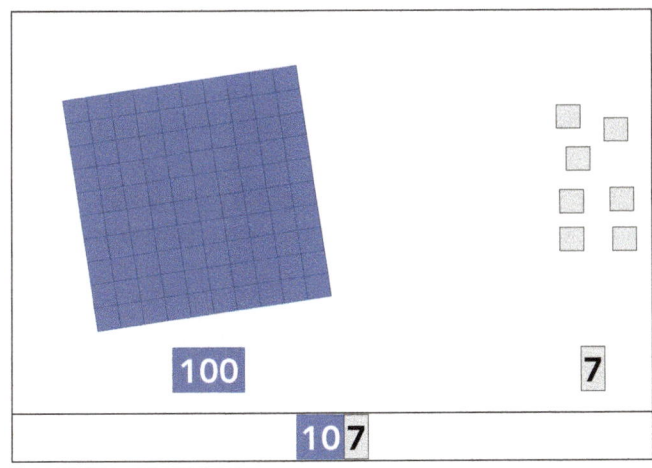

Source: Created by Robert Ronau.

In spite of the potential pitfalls, base 10 materials provide an excellent opportunity to model numbers and operations. Students can "see" the magnitude of a number, proportional relationships, and operations, tied to a physical representation that can help solidify students' understanding and acquisition of the algorithms.

Misconceptions that may be associated with a manipulative can be avoided by planning ahead and thinking through, and understanding the affordances and pitfalls that may occur when selecting and having students use any physical materials. When selecting any manipulative for your class, it is important to consider how well it represents the mathematical idea(s) you are teaching. Is there an explicit relationship between the manipulative and the mathematical concept? Is this relationship transparent? Are there any distracting features of the manipulative? Are there features of the manipulative or the use of the manipulative that might cause misconceptions to form?

Selecting the physical or concrete materials is one consideration, but how you introduce and use them in your classroom is another. Of the three representation types, many upper elementary students do not use physical materials as much as they should. But these materials lay the foundation for when these students become middle and high school students and begin to use materials such as Algebra Tiles™ to learn algebraic concepts and two-colored counters to represent integers. To make the use of concrete materials fully integrated into your instruction, it will be important for you to clearly tie their use to the mathematics content you expect your students to learn, so that they see how the manipulative will help them build their mathematical understanding. Additionally, seeing you use these thinking tools on a regular basis as you work through problems, as well as seeing other teachers in the school using them, will help students accept them and see their value as part of the mathematics learning process. Expect that it will take time for this use of concrete materials to feel natural for students to select what they need for use in their day-to-day mathematical learning experiences as part of the mathematical practices. However, keep working on it, and embrace the power of the C in CSA becoming a schoolwide effort through your MWSA team commitment. This will help establish these representations as a norm and expectation across all mathematics lessons. In addition, by doing so, you'll be setting your students up for further success in middle school, as the C in CSA is just as valuable to student learning even in high school.

As you are creating your MWSA, reflect on the physical materials you typically use, and make wise choices when selecting the materials you agree to have as part of your instruction. Are there other alternatives to the materials you have used in the past? Do you use these materials anyway, knowing that you need to provide explicit discussion about aspects that could cause misconceptions? What are other manipulatives that you would like to consider? These questions may help you get started in thinking through the systematic use of concrete representations by your MWSA team.

Semiconcrete and Abstract Representations

Equally important are the semiconcrete and abstract representations, which are the most common ones found in elementary curricular materials. Abstract representations, the symbols we use, are surprisingly not always well understood by our students. In fact, even at the high school level, more than 60% of the students surveyed felt that there are too many symbols to learn, use, and understand, which can cause them to not do well in mathematics (Chirume, 2012). Our choice of representations is important!

Semiconcrete representations provide a way for students to archive or keep track of the models they created with the physical materials and the actions they performed on them. Sketching or drawing supports the development of a mental residue of the actions students performed and provides a representation of the actions they can refer to later.

When combined with the abstract representations, students can see the connections across them and better remember the experience and the outcomes from using the materials. Figure 4.6 shows an example of a semiconcrete schema to use with additive word problems, such as the following:

Gina had eight stickers. Keenan gave her three more. How many stickers does Gina have now?

FIGURE 4.6 • SEMICONCRETE REPRESENTATIONS OF ADDITIVE WORD PROBLEMS

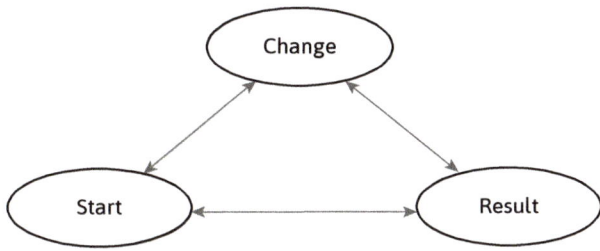

Notice that the schema depicts the possible relationships for result unknown, change unknown, and start unknown problems via a semiconcrete representation. Students need to use counters and talk about where the amounts can be placed on this graphic organizer to represent what Gina has and what she receives from Keenan. These

> **CORE MWSA IDEA**
>
> **Even commonly used mathematical symbols can affect student learning if students do not know their meaning!**

Mathematics sketch: A handmade representation of an idea; an image of a concrete representation or schema that is a rough version of the design or materials.

Mathematics drawing: An attempt to draw a careful representation of a geometric shape or figure, for example, with attention to being as precise and/or proportional as possible.

Mathematics illustration: A commercial image such as those from a textbook, computer program, or whiteboard display.

conversations paired with students' actions can help them solve the problem. Additionally, they should write the equation, or for the youngest students the teacher can write the equation initially, then allowing them to do so as soon as possible. As you are representing the mathematics, always use the symbol representations to reinforce the recording of the action.

Abstract representations also need some consideration, even though they are some of the most common representations we use. The way we write symbolic or abstract representations affects students' ability to use them optimally. For example, writing an addition problem horizontally rather than vertically creates a different perspective for students as it doesn't immediately suggest a solution strategy. This is a consideration for your MWSA. CSA provides a way to think about all the representations used in your lessons and how they fit within an instructional unit. What are some concrete materials that may enhance student learning? What are some semiconcrete representations that can be aligned with the concrete models? What are the characteristics of effective representations that support strong student learning of concepts? What are the appropriate abstract representations? How will these representations remain familiar as students progress up through the grades, and how do they build on students' previous mathematical learning experiences? Figure 4.7 provides some examples in each category of the CSA approach for you to consider.

FIGURE 4.7 • EXAMPLES OF CONCRETE, SEMICONCRETE, AND ABSTRACT REPRESENTATIONS

Concrete representations	Semiconcrete representations	Abstract representations
Counters	Drawings/sketches/illustrations	Numbers
Base 10 materials	Schemas/graphic organizers	Operation symbols
Color tiles	Graphs	Equal sign
Rulers	Bar or strip diagram	Inequality signs
Number balance	Place value mat	Words
Three- and two-dimensional shapes	Number lines	Tables
Fraction pieces	Hundreds chart	Letters as unknowns
Decimal squares	Grid paper	Ordered pairs
Percent necklace	Tables	
Two-color counters	Virtual manipulatives	
Cuisenaire Rods	Dynamic geometry software	
Connecting cubes		
Graphs made from physical materials		

While it may seem that these are small decisions, they collectively have a large impact on students' clarity, understanding, and success in their mathematics journey. The consistency across grades and courses provides students with a representational system that doesn't require them to learn something new, regardless of the mathematical focus or the teacher. It supports better communication and eliminates students guessing at what some unique representation means. Don't neglect discussing even what might appear to be the most basic symbols in your MWSA collaborative team meetings.

 **REFLECTION**

CONSTRUCTION ZONE–WHAT REPRESENTATIONS ARE MOST BENEFICIAL AND SPAN THE GRADES?

As you think about the representations you will use as part of your MWSA, consider these questions:

- Which representations can you agree on that will span multiple grades?
- Which representations have you used that are not productive in terms of helping students learn or for which you may not know all the options for using them?
- Which representations might cause confusion or create or perpetuate misconceptions?

Using the following space, record representations that are being used that need to be rethought, those that might need further explanation, and others that can and should be used across the grades. Then, as you continue reading this chapter, other suggestions may help you spark new ideas or prompt you to reconsider what can be used as appropriate alternatives.

COMMONLY USED REPRESENTATIONS IN ELEMENTARY SCHOOL

Let's explore some commonly used representations in elementary school mathematics that would benefit from some clarity and common agreements. Figure 4.8 provides a list of these representations, shares what needs to be considered, and provides appropriate alternatives grounded in research-informed and equitable instructional practices. This list does not include every representation, nor does it include every mathematics topic, but it should give you a starting point to consider other materials currently used in your particular school or district that might warrant revisiting.

Before we begin to look at the specifics, note the following practices that should be considered for your MWSA:

> **CORE MWSA IDEA**
>
> **Never avoid using representations because you think they "take too much time," as the investment is worth your energy and time for years to come.**

- Never avoid using representations because you think they "take too much time"—the investment is worth your energy and time for years to come. Remediation or intervention takes much longer!

- Teach students how to handle concrete manipulatives, including the etiquette of taking them out of the small bag or other container and putting them away. Model good and inappropriate approaches to handling the materials to explicitly define what you expect.

- Give students a moment to explore new manipulatives (including virtual manipulatives), but never say that you are giving them time to "play" with the manipulatives before you start the task. These representations are thinking tools, not toys.

- For concrete manipulatives, keep a basket on the table so that students can select the appropriate tools. The choices in the basket can change as the topics under investigation shift over the course of study.

- For semiconcrete representations, explicitly show students how to draw them. For example, show them how to draw a square, a line, and a dot to represent hundreds, tens, and ones, respectively, for base 10 materials. Also show them how to represent problems, so that a word problem with three monkeys in four trees is not a long-term project of drawing elaborate trees with funny monkeys. You can suggest that students instead represent the trees with a circle and each monkey with a dot as a "math sketch." This rough sketching keeps the focus on the mathematics—exactly where we want it!

FIGURE 4.8 • COMMONLY USED REPRESENTATIONS IN ELEMENTARY SCHOOL MATHEMATICS

Representations	Considerations	Suggested options
Concrete representations		
Counting tools for number sense and for developing cardinality	Young children learn to count discrete items, moving to eventually attaching a number to that quantity. Students eventually go on to counting groups of objects, such as tens (place value) and fours (multiplication).	There are unending options for counting whole numbers, from small plastic teddy bears to beans, to connecting cubes. The MWSA responsibility is to decide to regularly use them.
Various fraction representations	There are several representations to consider for fractions other than the common action of partitioning a region into parts. You need to consider all the possible concepts fractions can represent (with a brief example): *Part-whole*: shading a region or area *Division*: sharing equal-sized groups as a unit *Measurement*: identifying a length as a unit and indicating how many of that unit—or parts—determine an object's length *Operator*: considers a fraction as a multiple of a unit fraction *Ratio*: a comparison of two quantities as, for example, part-part or part-whole (ratio is commonly taught starting in sixth grade)	Three most common fraction representations are the following: 1. Regional or area model (e.g., circular fraction piece, pattern block, tangram, geoboard) 2. Length model (e.g., walk on number lines, Cuisenaire Rods, rulers, paper strips) 3. Set or discrete model (e.g., two-color counters), or objects related to the context (e.g., crayons, small toy cars) When comparing the size of two fractions, when possible, use a model such as a number line or concrete fraction pieces (e.g., circles, rectangles, fraction strips) that are premade.
Arrays	The array, or rectangular arrangement of rows and columns, is first used as an essential representation for multiplication. It can later be used for area, for multiplication of fractions, and in secondary courses for matrix algebra.	Base 10 materials can be paired with arrays to build place value relationships while modeling the operations. For example, when students are learning multidigit multiplication (24 × 32), there is a powerful connection between making an array with base 10 materials that has dimensions of 24 by 32 and locating the partial products through the consideration of the distributive property. This action enables students to segue to an open-array, semiconcrete representation over time.

(continued . . .)

(continued . . .)

Representations	Considerations	Suggested options
Measuring tools	The measuring tools that are selected have an impact on student learning. They may affect the level of accuracy and precision as well as the interpretations that students make about the measure. Rulers can be challenging, depending on whether or not the ruler has a leading zero. If there is no leading zero, students are often confused about where to start when they are measuring length, especially if the end of the ruler is represented as chipped or broken. The protractor as a tool for measuring angles is one that is not intuitive, given the construction of the tool itself.	There are other options for measuring tools that can be explored to better support student learning. For length, students should start with using multiple 1-in. or 1-cm tiles to show how many units are needed to match the length (equal partitioning). For longer lengths, they should experience a trundle wheel. For older students, for example, the angle ruler (or goniometer) may be more easily read and interpreted than a protractor (check your local requirements to see if protractors are the only tool allowed on assessments). Also, explore using square units that students can iterate to cover an area, and filling shapes with cubes for a way to explore volume.
Geometric shapes and their attributes	You can move students to interpret the properties of shapes; if you engage students in a task where they explore the number of ways in which five squares can be placed, they will notice patterns related to orientation and changes in perimeter based on the configurations. They may also observe that some of these figures that are called pentominoes (nets) can be folded to make open-top boxes while others cannot. This will lead to conjecturing about the characteristics of the configurations that limit the folding.	There are many materials for teaching geometry that are appropriate, including pentominoes, tangrams, two-dimensional shapes, geometric solids, geoboards, loops of yarn, skeletons of shapes with stirrers and clay balls, digital options, etc. Using the illustrations in books alone to teach geometry results in many missed opportunities for students to enhance their geometric experiences and develop a love for this very visual mathematical domain. In your MWSA, decide how the same tools will be used in different grades.

Representations	Considerations	Suggested options
Coordinate graphs	We often assume that students understand grid paper as a means of location. But using our CSA approach starting with grid paper as a representation is not enough. Students must walk on the grid and locate small objects through coordinates first. Give students cards with coordinates for them to walk to, and then put the card down on the correct intersection. Or place small objects at locations, and have them record the ordered pairs for each.	Use a large tarp or painter's drop cloth, and mark it with intervals as a grid where your students can easily stand. You can also paint a large grid on the playground or parking lot to serve the same purpose. Then the grid paper becomes a logical way to record what they experienced.

Semiconcrete representations

Representations	Considerations	Suggested options
Number lines	Number lines are especially appropriate for specific contexts. For example, when considering a length context such as locating a runner in a race to see how far they've run, students should consider the number line representation. But students can also use a number line to represent a variety of concepts such as number magnitude and number operations; vertical number lines are used as scales for temperature and capacity, for example. Perpendicular number lines form the coordinate grid that comes to play in fifth grade. Number lines are one of the most important representations at the elementary grade level. Students can use double number lines either in print or virtually to compare fractions or decimal values. Remember that challenges will arise with interpreting student-partitioned number lines as they have great difficulty making equal intervals.	Although we include number lines under semiconcrete representations, they of course start with concrete versions—with number paths, walkable floor number lines, and clothesline number lines; then to traditional tick-marked lines; and later to more abstract versions via empty number lines. Initial confusion may occur if you do not make explicit in the early grades the connections between the units in a number path, for example, and the units of length as shown in a continuous line.
Bar, strip, and tape diagrams	Bar, strip, and tape diagrams are especially helpful to engage students in both additive and multiplicative thinking by showing number relationships. In each of these representations, labeling each bar, tape, or strip is important (Van de Walle et al., 2019). They will also link effectively to middle school explorations of ratio and proportions.	Although we include bar, tape, and strip diagrams under semiconcrete representations, concrete versions of the part-part-whole diagram or comparison multiplication problems can also be used as a precursor.

(continued . . .)

(continued . . .)

Representations	Considerations	Suggested options
Geometric shapes	The way in which geometric shapes are drawn (note that we are not saying *sketches* here) has an effect on students' ability to interact with the mathematical ideas embedded in the representation. These geometrical drawings occur in most kindergarten standards. Even though it is the case that we cannot visually rely on accurate drawings and must use other information to make the drawing more accurate, having drawings deviate too far from the intent`` can be problematic.	As an MWSA team, decide what degree of precision you will use for geometric drawings. Are students always expected to use a straightedge? In what grade? Using your standards, determine, for example, how in fourth grade you will have students represent and notate parallel lines, perpendicular lines, obtuse angles, and so on.
Line segments drawn instead of lines	The language we use and the representations should be aligned so that the representations match the way we talk or write about them. For example, when a number line is drawn, it is sometimes drawn without the arrows at the end, needed to indicate that it is a line. By fourth grade, we may think that students "know" that it is a line, but for many students who struggle with this idea, the representation of a line segment versus a line makes a difference.	As your MWSA discussions continue, determine how lines, line segments, and rays will be represented. Be consistent in the representations so that students also become more consistent and accurate in their drawings.
Abstract representations		
Equations written in different formats	Many students will see, for example, the equation $4 + 5 = 9$ as the only way to write the equation. They might suggest that $9 = 4 + 5$ is backward, $6 = 6$ makes no sense as there is nothing to "do," and $3 + ? = 9 + 4$ is just plain confusing. In that case, students often add all of the numbers as the answer to the unknown. Some of this confusion is tied to their understanding of the equal sign, but consistently using these different formats encourages the dispelling of those misconceptions.	Write the equation that best models the situation from the context, starting from kindergarten on. So if students are asked to put 10 goldfish in two tanks in as many ways as possible, start them off with recording their findings with $10 =$ to link the symbolic representation directly to the scenario. Students need to experience all these different equation formats early on in order to recognize the connection of the context to the symbolic representations.

Representations	Considerations	Suggested options
Unknowns written in equations	In elementary school, unknowns are often represented by blank spaces, such as 12 + 34 = . This approach is not the best in having students make connections across the grades.	Initially, you may want to start by using an empty box to represent an unknown, then move to a question mark and then to the use of letters in upper elementary. Consider your local curriculum and assessments as you make your MWSA decisions on this important topic. Write the letters that you are using as unknowns in equations or as variables in *italics* so that they are distinguishable from the text and operation symbols. Use *y* rather than y. Avoid using letters that align to words in the problem, such as using *c* = cakes. *c* is not a label representing cakes; it is a quantity, the *number* of cakes. This helps students avoid thinking that 4*c* is "four cakes." If you are handwriting variables, use cursive rather than manuscript to show the italics.
Operation symbols	When students first use multiplication, for example, the symbol × is used to denote that operation, but later there are more signals for multiplication, including exponents. Division has multiple signs shown from the start, but ensure that you show that a fraction is seen as division too.	Make sure you use a variety of symbols to represent the operations. Consult your local materials and assessments to see what should be put into your MWSA and linked effectively to concrete and semiconcrete representations.
Rational numbers	Often rational numbers are written as 2/3. This is because the expression is difficult to interpret. See the discussion in Chapter 3. With decimals, writing a decimal as .78 does not adequately represent the place value connotations as compared with writing it as 0.78. Visually, it is much easier for students to "see" the decimal representation and not confuse the decimal point with a period or other text representation.	Use the vinculum (horizontal fraction bar) to write all rational numbers that are in fraction form. This clearly delineates the numerator and the denominator, $\frac{2}{3}$ rather than 2/3. Write decimals with a leading zero to support students' understanding that, for example, 0.35 indicates a number less than 1 because there is no *whole*.

Note: The representations presented in this table are synthesized from those that we have collected from educators as well as based on our previous work in Karp et al. (2014, 2015) and Dougherty, Bush, and Karp (2017).

Remember, merely handing materials to students for them to use to represent ideas, without sufficient exploration of the material itself, may cause students to miss some of the most insightful parts of mathematical investigations. As you plan to use concrete materials in particular, think about the opportunities that may present themselves when students create with the manipulative. Reevaluate all tasks in which a variety of representations could be used for concept development or problem-solving, and consider changing tasks (which you will learn in Chapter 6) to offer more opportunities to do so.

Now that you've examined some of the representations that are commonly found in elementary school, how they can strengthen understanding, and suggested alternatives, revisit your Representation Wall. Some of the representations you listed might be included in our commonly used list, and others might be unique to your setting. Use the template below to brainstorm with your team the next steps you will take to identify appropriate representations for use in your grade and how they will be linked to earlier and later grades.

THINGS TO DO

Representations MWSA sample table

Our currently used representations	Agreed-on representations

 Available for download at resources.corwin.com/mathpact-elementary

SHARING WITH STAKEHOLDERS

As we've been sharing throughout this book, a key component of the MWSA is ensuring that every stakeholder in your elementary school students' mathematical learning process is informed and on board. Now, you'll work to develop some communication tools that will inform everyone in your MWSA community to work together to use appropriate representations.

⇨ TRY IT OUT ⇦

MWSA HANDOUT FOR REPRESENTATIONS

Representations We Are Using in _____

Representations that may cause confusion	Agreed-on representations in our whole school agreement

online resources ⤵ Available for download in English and Spanish at **resources.corwin.com/mathpact-elementary**

As you let families know about the work you are doing, here is a letter that can be shared. This message will help onboard parents and families or others who might be working with your students, including tutoring services. We suggest that all grade-level teachers and others in the building who are engaged in the teaching of mathematics sign the document.

THINGS TO DO

Send the Letter

Hello _____,

We have already written to you about the Mathematics Whole School Agreement (MWSA) that we are developing across the entire school this year. As you know, we are all working hard to align our instruction in mathematics across the grades. As you may remember, earlier this year you received a letter where we talked about the mathematical language and notation we use during instruction. We are now looking at the representations we use in mathematics. As a mathematics team, we have agreed on the physical materials we may use to model the mathematics and the ways in which we explain the mathematics by means of pictures or diagrams and mathematical symbols. Everyone in the school involved in the teaching and learning of mathematics is using these and is focused on teaching for students' depth of understanding and connection to mathematical ideas within and across grades. The way we model in mathematics has an effect on the way students understand mathematical ideas. We want your student to become an adult who knows mathematics and will succeed in whatever they choose to do in life. We thank you for joining us in making this shift to be consistent in how we support your student as we prepare them for their personal and professional future.

Thank you for your help,

Your student's teachers and principal and members of the school community

---

 Available for download in English and Spanish at **resources.corwin.com/mathpact-elementary**

PUTTING IT ALL TOGETHER!

At this point in your discussions, it is important to think about the rationale for why you need to understand the reasons for using consistent and appropriate representations. As you consider these reasons through the overall lens of the MWSA and advocate for an MWSA in your school or district, here are some key talking points for being consistent in the ways in which we model mathematical ideas:

- Using consistent representations helps students connect mathematical ideas.
- Incorporating the use of CSA throughout the instructional units builds students' conceptual understanding and mental residue.
- Appropriate representations help students communicate their thinking more clearly and more accurately.
- Accurate representations lead to a deeper understanding of the mathematical ideas.

NEXT STEPS

In Chapter 3 you examined the importance of the precise use of mathematical notation, and those ideas were continued in this chapter. Up to this point, you have considered mathematical language such as vocabulary, notation, and concrete, semiconcrete, and abstract representations as part of your MWSA.

In Chapter 5 you will continue the development of your MWSA as you consider rules that expire, their effect on student learning, and how they may actually interfere with their understanding. Students may bring some rules with them to school from home or from other schools they attended, and we may use other rules in our classes that we assume "help" students learn. You will need to consider how to assist students in determining which rules have expired and to determine which rules are generalizable.

WHY WAS I TAUGHT THAT?

Evaluating Rules That Expire

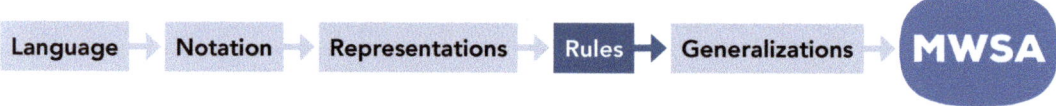

When we solve an equation, why does the answer need to be on the right side of the equal sign? Can't think of a reason? That's because no mathematical reason exists that mandates the placement of the answer there. But it makes you better understand why students are uncomfortable when they are asked to solve the following:

☐ + 2 = 7
8 = ☐ + 3
2 + 4 = ☐ +5

For many students, these equations don't look correct, and they will even suggest that the equation 8 = ? + 3 is written backward!

Why do you invert and multiply when dividing fractions? Not sure? Let's be honest: It is likely because you learned this "trick"—just as we authors did as students (and taught as teachers)—but you possibly weren't taught conceptually what was happening mathematically when dividing fractions using this algorithm. Tricks, shortcuts, or rules such as the two mentioned here are what we describe as "rules that expire," or RTEs, and eliminating them and trading them for more meaningful approaches is a key component in the creation of a strong and useful MWSA.

In this chapter you will learn

- What qualifies as an RTE
- Why RTEs negatively affect students
- Examples of commonly used RTEs and suggested instructional alternatives

Remember, as you work through this chapter, you're actively establishing the RTE component of your MWSA—you're making great progress!

WHAT ARE RTEs?

RTEs are a deeply rooted tradition in mathematics education, a means to teach a procedure or a strategy in a way that the teacher believes makes the learning easy and fast or helps students remember. Sometimes RTEs are used with the best of intentions as an attempt to make learning "fun." However, let's be clear: RTEs are harmful in the long term and should not be used. We authors learned this the hard way by teaching these rules in our classrooms only to regret it later when we taught other grades or learned more mathematics content. RTEs might temporarily seem to help in the short run, but in the long run they support the myth that mathematics is a set of disconnected tricks and shortcuts, is magical, or at worst is incomprehensible. The basic premise of RTEs is to teach for convenience or speed, and the subsequent initial appearance of student success fuels the continuance of teaching these rules. In other words, being able to apply RTEs by rote may get students through the next problem, quiz, test, or high-stakes assessment, making it seem as though there is deep conceptual understanding (or a strong reason to teach this way) when often there is not. Then, when that appearance of success leads us to believe that students understand more than they do, we use the RTEs again. In essence, the use of the "trick" or the "shortcut" becomes a self-fulfilling prophecy. Instead, we should teach for the future mathematics we know is coming and emphasize enduring understanding and long-term utility. Instruction that fosters students' depth of understanding builds procedural fluency *from* conceptual understanding (as described in NCTM, 2014b). Smith et al. (2017) state,

> **Rules that expire:** Tricks, shortcuts, or rules that are used in mathematics that immediately or later fall apart or do not promote mathematical understanding.

CORE MWSA IDEA

Even actions we take as teachers that seem well-meaning can be harmful in the long run!

CORE MWSA IDEA

Teaching for understanding and long-term utility prepares students to become adults who are mathematically literate.

> Throughout their mathematical experiences, students should be able to select procedures that are appropriate for a mathematical situation, implement those procedures effectively and efficiently, and reflect on the result in meaningful ways. This procedural fluency, however, is fragile and meaningless without a sound conceptual understanding of the mathematics. Conceptual understanding and procedural fluency are essential and integrated components of mathematical proficiency. (p. 55)

Teaching using RTEs equates to teaching in a way that is often devoid of sound conceptual understanding, which leads to students using procedures in ways that are meaningless and can conceal fragile and often incomplete knowledge. This approach is the opposite of what our students need and deserve.

Eradicating the teaching of RTEs is a key commitment of the MWSA. In fact, the RTEs movement, first published by the NCTM in articles in the three main practitioner journals at the time—*Teaching Children Mathematics* (Karp et al., 2014), *Mathematics Teaching in the Middle School* (Karp et al., 2015), and *Mathematics Teacher* (Dougherty, Bush, & Karp, 2017)—served as the catalyst for our MWSA work and ultimately the writing of this book. The case to avoid RTEs is also shared by kindred spirits, such as in *Nix the Tricks* by Cardone and MTBoS (2015). But remember that the reason we know so many of these rules is because we used to teach them ourselves. We (the authors) have all been there; we've all taught in these ways. However, we now know better, and once we knew better, we knew we could do better. Similarly, once any teacher knows better, Leinwand (1994) suggests that there is a professional obligation to do better.

CORE MWSA IDEA

Don't dwell on the past, but now that you know better, do better!

Examples of RTEs

Multiplication makes a number bigger. This is a rule sometimes taught in the elementary grades that can clearly cause confusion in middle school and beyond. When students are first learning to multiply whole numbers, multiplication is often tied to whole-number addition, so it would appear to make sense to tell students that multiplication makes a number bigger—meaning anytime they multiply two numbers the product is a value greater than either of the two numbers being multiplied. Just as the comparable rule that *addition makes bigger and multiplication makes bigger* seems at the time to be harmless and even helpful to students, likewise students might similarly be taught that *subtraction makes smaller and division makes smaller*. These rules fell apart the moment they were taught—even with whole numbers (e.g., $1 \times 0 = 0$). They continue to expire when we multiply or divide fractions or when—starting in the middle grades—students add, subtract, multiply, or divide integers. This limits students' ability to reason with generalized quantities in tasks. In addition, these RTEs foster the use of less desirable mathematical language, such as *makes*, which we know from Chapter 2 is not the best word we can use to describe equality, as well as the comparative words *bigger* or *smaller* to describe quantities when the preferred mathematical language includes words such as *greater than*, *less than*, and *fewer*.

The relationship of the operations to the result of the operation carries over even through middle school and into high school and affects the way students think about generalized quantities. Consider this problem given on a conceptual progress monitoring assessment (Dougherty, Foegen, & DeLeeuw, 2017):

Mari said, "$2h$ is always greater than $h + 2$." Do you agree with Mari?

A. Disagree, because it is possible that $2h$ can be equal to or less than $h + 2$.

B. Disagree, because multiplication is not the inverse of addition.

C. Agree, because multiplication always gives you a larger answer than addition.

D. Agree, because h is a positive number.

At the end of the second-semester Algebra I in high school, out of the 750 students who responded to that item, 41.6% selected "C" as the correct answer. This choice clearly indicates that a large number of students still believe that multiplication makes numbers larger. It is also surprising to note that only 35.5% selected "A," the correct answer, in spite of having completed a formal algebra course.

Use keywords to solve word problems. This is an RTE that is widespread and deserves much thought and attention. Keywords are frequently introduced in the elementary grades as an important strategy for solving word problems, especially for students who are currently struggling or are emergent bilinguals. Students are often told that when they see specific keywords in a word problem, those words are signals that indicate which operation to perform on the numbers. Suppose students are given the following problem:

Janette has 8 apples. Roberto has 3 apples. How many apples do Janette and Roberto have altogether?

With a focus solely on a keyword approach, students would identify *altogether* as the keyword signaling them to add the two numbers (8 and 3) and would simply solve the word problem by stating $8 + 3 = 11$. While a keyword approach seems to work for this problem, there are five important reasons for avoiding this approach (as described in Karp et al., 2019):

1. It removes all need for students to make sense of the problem presented and the mathematics included in it.

2. Students become incorrectly influenced by keywords taken out of context.

3. Many problems simply do not contain keywords, which will leave students at a loss if they rely on this strategy.

4. The strategy of using keywords completely falls apart as soon as students encounter multistep problems.

5. When using this strategy, students are not gaining practice in comprehending mathematical situations and they are not applying prior knowledge.

Figure 5.1 showcases some sample keywords, which are often displayed on posters in a classroom to jog students' memories. Notice that these posters relate to similar ones we shared in Chapter 1, where underlining words was a step in several mnemonics. In addition to the reasons we just listed, many keywords can actually link to multiple operations, and without making sense of the problem; for example, students might simply add because they see the word *altogether*, but the scenario is actually calling for subtraction or multiplication.

FIGURE 5.1 • EXAMPLE OF A KEYWORDS POSTER

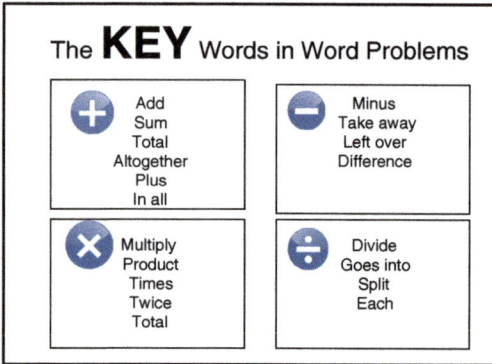 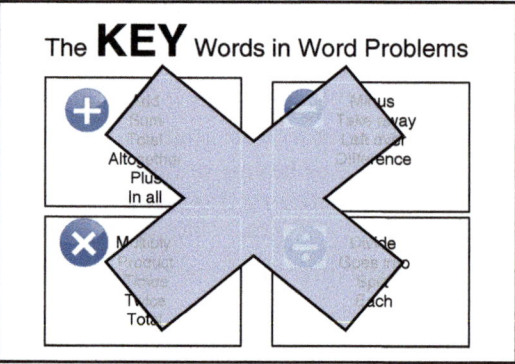

Source: Kobett and Karp (2020).

If you are still hesitant to stop using a keyword approach, here are two examples where keywords clearly do not work:

Dominique has 4 boxes of colored pencils. There are 12 colored pencils in each box. How many colored pencils does she have in all?

Here, students might see *in all* and incorrectly add 4 and 12, not recognizing that the context of the problem indicates a multiplication situation (4 groups of 12).

Daniella and her 3 friends are playing a game that has 40 cards. To start the game, Daniella needs to give out all of the cards, making sure that she and her 3 friends all have the same number of cards. How many cards will a player start with?

In this example, *all*, a word that is attached to addition in a keywords approach, is used twice but not related to a mathematical operation. Additionally, the phrase *how many* is used, which could also prompt students to incorrectly add the numbers. However, this word problem actually calls for division to be used, and there are no commonly used keywords that signal division in this word problem. Furthermore, without reading and making sense of the problem, students are likely to incorrectly use 3, rather than 4 (the total number of friends playing the game), in any operation performed.

It is clear to see, just from these two examples, how the RTE "Use keywords to solve word problems" expired the moment it was introduced! An important agreement for an MWSA is to stop using a keyword approach to solve word problems and to banish all keyword posters in the school!

There are many other, better ways to engage students in solving word problems (see Figure 5.2). We list these approaches here. (For a more in-depth discussion, see Karp et al., 2019.)

FIGURE 5.2 • APPROACHES TO SUPPORT STUDENTS IN SOLVING WORD PROBLEMS

Approaches	Descriptions
Increase readability	This includes ease of reading, length of the problem, and elimination of the use of second-person pronouns to address the student
Ensure topic relevance	Choose or adapt word problems so that the context is relevant to your students (which is not the same as being relevant to you as the teacher)
Support the reading and understanding of mathematical words	Chapter 2 provides guidance on this topic
Use concrete materials	Chapter 4 provides guidance on this topic
Have students imagine the situation	Get students imagining, articulating, and referring back to the context of the problem
Employ schema-based instruction	Employ schema-based instruction: Have students carry out the actions in the word problem (see Chapter 4 for more information)
Have students paraphrase the problem in their own words	Students can paraphrase the word problem in their own words to help ensure that they comprehend the context
Have students work backward to see how word problems are generated	Students can start by sketching a picture or writing an equation, for example, and then create a word problem for that representation
Act out the problem	Have students act out the word problem; this can be helpful for students of all ages but particularly for those students still learning to read, write, and draw mathematical representations
Include an advanced organizer	Consider moving the question being asked in the word problem to the beginning of the problem
Give students a problem without a solution or without numbers	This practice helps students focus on the context of the word problem rather than trying to quickly perform an operation with the numbers
Give students a problem without a question	Have students consider what mathematical questions could be asked given a situation, and then have students solve them

Negative Impacts of Teaching RTEs

There are multiple reasons why RTEs should be eradicated. First, as we mentioned, RTEs are not mathematically sound or correct—they often crumble immediately and by definition will fall apart later in students' mathematics careers. Second, RTEs perpetuate the negative perception and incorrect myth of mathematics as a set of disjointed ideas that are not connected to our world. These rules cause students confusion as they move vertically through the grades, when they start to believe that their teachers are teaching them "different" and sometimes conflicting mathematics each year. Depth and coherence of ideas within and across grades are thereby hindered.

CORE MWSA IDEA

RTEs undermine the joy, beauty, and wonder of the discipline of mathematics!

A third reason for not teaching RTEs is grounded in the idea that mathematics can and should be so much more than the type of mathematics that exists in a space where RTEs are commonplace. Mathematics should be about learning big ideas and making a network of mental connections. Students should feel empowered and inspired by mathematics, recognizing that mathematics is essential and useful for making sense of our world.

You might meet this needed change with some initial skepticism. It can be hard to move away from teaching in ways that we have used for years and from how we ourselves were taught as students. "My students can't understand higher-level math, but they do great with this trick" and "This is all well and good, but we don't have enough time!" are two statements Cardone and MTBoS identified (2015, pp. 2–3) that you might hear from colleagues, although we hope you do not! Simply put, if students only know tricks, they are only doing tricks and not doing mathematics. And yes, while universally in education there is never enough time to do what we want, this statement is basically saying that we don't have time to actually teach mathematical ideas via the required mathematical practices and processes. We're sure you'd agree that is unreasonable. You may also hear well-meaning comments such as "Well, I tell my students this rule only applies to this situation," "They'll just understand that this rule doesn't apply when they get to high school," or "I always tell them they can only use the rule for _____." However, these comments and practices are also concerning because we know from our years of experience that students do not "grow out" of rules and the long-term effects of teaching RTEs are very real. For example, a study by Dougherty and Foegen (2014) found that when 652 second-semester, ninth-grade, high school Algebra I students were asked to identify an equation whose solution was 3, 31% of them selected $x + 2 = 3$ because they understood that the answer comes after the equal sign. These students still believed the RTEs that they were likely taught more than 5 or 6 years earlier; hence, they did not "grow out" of these rules.

Impacts of teaching RTEs that are important to remember include the following:

- Students use rules as they have interpreted them (often not as you tell them to interpret them).
- Students do not often think of the rule beyond its immediate application (even if you provide a caveat).
- Even those students you perceive as your strongest mathematics students find it unnerving and scary when they realize that what they believed to be fundamentally true as a "rule" no longer works.
- Once rules are taught, they are challenging to unteach. Why would we want to teach something that students just need to unlearn later?

Now it's time for you to reflect on the RTEs that you may currently use in mathematics instruction, that you learned as a young student, that you have used in the past, that students arrive at your class believing as truths, or that you've seen in your curriculum resources or around your school. Consider this as a starting point for discussion with your teaching partner, grade-level learning community, or school mathematics team as you begin creating your commitments to avoid RTEs today!

CORE MWSA IDEA

Come to an agreement as a team, and commit to stop using RTEs.

CONSTRUCTION ZONE– RTE ERADICATION WALL

As unnerving as it may feel, take some time to reflect on and discuss the following:

- What RTEs need to be eradicated from your classroom, grade, or school?
- What RTEs have you taught students?
- Which RTEs have you learned yourself and now realize are bound to confuse students, even though you may have survived them?

In the space that follows, jot down the RTEs that you have heard of or that may be used in your setting that need to be eradicated. Come to an agreement as a team, and commit to stop using these RTEs. Then, as you continue reading this chapter, we'll provide suggestions for replacing these RTEs with more effective instructional strategies that meet students' long-term needs.

COMMONLY USED RTEs IN THE ELEMENTARY GRADES

In this section, we share some RTEs commonly used in the elementary grades. Figure 5.3 provides a list of RTEs frequently heard in the early years of schooling, explains why the rules expire, and provides suggested instructional alternatives grounded in research-informed and equitable instructional practices. This list does not include every rule, as we simply do not have the space to list every possible commonly used RTE and many more are locally based or invented and used by individual teachers or schools. But we believe that if we share these, you will rapidly become RTE detectives and sniff out others not mentioned here.

CORE MWSA IDEA

It's our job to collectively become RTE detectives!

FIGURE 5.3 • RULES THAT EXPIRE COMMONLY USED IN ELEMENTARY SCHOOL AND SUGGESTED ALTERNATIVES

Rule that expires	Expiration details	Suggested alternatives
Numbers and operations and algebraic thinking		
Addition makes numbers bigger, or you should always expect a larger answer when you add.	When students begin learning about the operations of addition, they are often given this rule as a means to develop a generalization relative to operation sense. However, the rule has many counterexamples—some immediate. For example, addition with zero does not generate a sum larger than either addend. It is also untrue when adding two negative numbers (e.g., –3 + –2 = –5), because –5 is less than both addends.	The main focus should be on teaching that the meaning of addition involves combining or joining quantities. Students should talk about the reasonableness of their answers, and giving them "fake" student work where mistakes can be discussed is far better than just giving them a rule that "bigger" answers are expected and that condition alone means that they have reasonable answers.
You cannot take away a larger number from a smaller number.	Students may first hear this phrase as they learn to subtract whole numbers. When students are restricted to only the set of whole numbers, subtracting a larger number from a smaller one results in a negative number—an integer that is not in the set of whole numbers—so this rule is true. But later, when students encounter applications or word problems involving contexts that include integers, they learn that this "rule" is not true for all problems.	In the early years, this rule can cause a great deal of confusion. When students are subtracting two-digit numbers as they think of this rule they might give this response: $$\begin{array}{r} 32 \\ -\,15 \\ \hline 23 \\ \hline \end{array}$$ The preferred way to present this idea is to use concrete materials—such as base 10 materials for the double-digit subtraction problem above. Start by representing the minuend with the materials; so in this case you would show 3 tens and 2 ones (preferably on a place value mat). Ask a student to remove 5 ones. Hopefully, they will not say you cannot take away a larger number from a smaller and instead say, "I can't do that." Ask them to consider the entire quantity of 32—can you take away 5 from 32? Hopefully, they will agree that that is possible. Then you work through the process of regrouping. Key to this process is the conversation about the magnitude of the minuend (not a single digit) and continually connecting back to place value representations to build understanding.

(continued . . .)

(continued . . .)

Rule that expires	Expiration details	Suggested alternatives
Multiplication makes numbers bigger, or you should always expect a larger answer when you multiply.	In the case of the equation below, the product is smaller than either of the factors: $$\frac{2}{5} \times \frac{1}{3} = \frac{2}{15}$$ This relationship is also the case when one of the factors is a negative number and the other factor is positive, such as $-3 \times 8 = -24$.	Attention should be given to multiplication as a way to consider equal groups or comparisons using a scale or rate. Focusing on the meaning of the operation and the properties of multiplication is the best approach.
Multiplication is repeated addition.	While multiplication can be thought of and written as a repeated addition equation, when students only think of multiplication in this way, they might overgeneralize this idea. For example, they might believe that 4^3 is $4 + 4 + 4$, instead of $4 \times 4 \times 4$. Or they may think that you can use repeated addition for fraction factors, such as in the problem $$\frac{1}{4} \times \frac{1}{3} = ?$$	Write expressions, such as 4^3 and others, in expanded form to reinforce the meaning and help counteract any misunderstanding. Some students will try to add 18, 23 times to find the product of 23×18. This is inefficient and opens up more opportunity for error.
When you multiply a number by 10, just put a 0 on the end of the number.	This rule is commonly first taught when students are learning to multiply a whole number by 10. But it becomes routine and is practiced as students multiply 15×10 or 241×10. However, this pattern is not true when multiplying decimals (e.g., $0.46 \times 10 = 4.6$, not 0.460). This "rule" may reflect a regular pattern with whole numbers, but it is not generalizable to other sets of numbers.	Again, the key here is to help students reason and develop the number sense to know when their answers are reasonable. As with many of the other RTEs, it is best just not to teach this rule. What is preferred is to let students identify this pattern if they notice it, and then you can treat their conjecture just like others that students make. You first test it with a variety of numbers. Then you can create the needed boundaries by asking them if this pattern will work with nonexamples, such as $\frac{1}{2}$ multiplied by 10.

Rule that expires	Expiration details	Suggested alternatives
You always divide the larger number by the smaller number.	This rule may seem always true when students begin to learn their basic facts for whole-number division and the computations are not contextually based. But, for example, if the problem reads, "Fashid has 2 cookies to divide among himself and two friends. What is the portion for each person?" The answer is "$2 \div 3$." Another example would be a problem in which one number is a fraction: Gordy has $\frac{1}{2}$ of his birthday cake left. He wants to share it with 3 friends. What portion of the whole cake will his 3 friends get? $$\frac{1}{2} \div 3 = \frac{1}{6}$$	The focus needs to be on the meaning of division as a way to partition or share equal groups. Using concrete materials or sketches can help support students' thinking about the division process.
To factor, use a factor rainbow.	For factoring numbers, students are sometimes told to create a "factor rainbow" listing all of the factors of a number from the least to the greatest. For example, for 30, a student would first write 1 and 30, then 2 and 15, then 3 and 10, and then 5 and 6. A factor rainbow is intended to help students list all of the factors, knowing that you are finished when the center of your rainbow has two consecutive (or in some cases the same) numbers. However, factors can easily be missed with this rule, and many numbers do not have factors that are consecutive numbers (e.g., 10, 36, and 50). This rule also expires when factoring square numbers that do not have pairs of two unique factors or when working with nonwhole numbers, such as work with variables and fractions in algebra.	Keep the focus on developing students' multiplication fluency and flexibility. When students truly have fluency, rather than just memorized facts, it is easier for them to use their own strategies to determine the factors of a number. Instead, have students capitalize on knowing divisibility rules. This approach supports number sense. While on the surface the rainbow appears to be a good visual structure, for the reasons we have stated it likely could do more harm than good.

(continued . . .)

(continued . . .)

Rule that expires	Expiration details	Suggested alternatives
Use keywords to solve word problems.	A keyword approach is frequently introduced in the elementary grades as an important strategy for solving word problems. However, as described earlier in this chapter, there are many reasons for avoiding this approach.	Earlier in this chapter we provided a list of many other, more meaningful and effective approaches that can be employed to help students build understanding of how to solve word problems.
Equations should be written so that the problem is always to the left of the equal sign: $9 - 6 = \square$.	When solving equations, students are often told to write their final solution in the form $5 \times 7 = \square$. This rule can confuse the meaning of the equal sign (Dougherty, DeLeeuw, & Foegen, 2017), as the equal sign actually indicates that the two quantities on either side are equivalent. Forcing a solution to be written in the form $x = \square$ detracts from the meaning of the equal sign.	There are equivalent forms of any equation, with instruction beginning in the primary grades. Students should also be encouraged to use equivalent equations as appropriate to a problem and its context. They should have ample opportunities to experience multiple equation forms, including the following: $17 = \square + 4$ $3 + 9 = \square + 5$
To compute accurately, use PEMDAS—Please Excuse My Dear Aunt Sally.	When students are learning about the order of operations when simplifying multistep numerical expressions, the mnemonic PEMDAS (or BEMDAS/BODMAS, as discussed in Chapter 2) is sometimes taught. As discussed in Chapter 2, PEMDAS is problematic because it encourages overgeneralizations and discourages flexibility in thinking.	We suggest focusing on students making sense of the problem and encouraging flexible thinking. However, if you are using a hierarchical model, do not use PEMDAS (or BEMDAS/BODMAS) but consider this order instead: a. Grouping symbols b. Exponents c. Multiplication or division d. Addition or subtraction
Fractions, decimals, and percentages		
To divide fractions, use the KFC method: Keep-Flip-Change.	When dividing fractions, students are sometimes told to Keep the first fraction the same, Flip the second fraction (invert the numerator and denominator), and Change the division operation to multiplication. Although this rule works mathematically, it's troublesome for several reasons. First, students overgeneralize it to other fraction operations. Second, this rule does not promote conceptual understanding or procedural fluency. Third, it does not support students as they try to determine the reasonableness of their answer.	The division of fractions should be connected to whole-number division by asking how many groups of the divisor make up the dividend. Although students will eventually be taught to use the algorithm, conceptual understanding should be developed first through the use of physical models (e.g., area, length; Cramer et al., 2010), through explaining the use of partitioning and sharing, or other methods such as the common-denominator strategy (Van de Walle et al., 2019).

Rule that expires	Expiration details	Suggested alternatives
"Improper fractions" should always be written as a mixed number.	When students are first learning about fractions, they are often taught to always change "improper fractions" to mixed numbers, perhaps so that they can better visualize how many *wholes* and *parts* the number represents. This rule can certainly help students understand that positive mixed numbers can represent a value greater than one whole, but it can be troublesome when students are working within a specific mathematical context or real-world situation that requires them to use fractions whose value is greater than or equal to 1. This frequently occurs first when students begin using fractions greater than or equal to 1 to compute and again when students later learn about the slope of a line and must represent the slope as the $\frac{rise}{run}$, which is sometimes appropriately and usefully expressed in this fashion.	Students need to be flexible in how they represent a fraction greater than 1 (or equal to 1). A better approach is to consider using a clothesline as an extended number line for fractions to be positioned. Give students cards that include matches of equivalent fractions written as mixed numbers or fractions greater than or equal to 1. Have them consistently see that they locate these pairs at the same place on the clothesline, and it doesn't matter how they are written. What matters is choosing which symbolic representation should be used in the context of the problem.
The longer the number, the larger the number.	When working with multidigit whole numbers that differ in the number of digits, the length of the number does reflect the magnitude. However, it is particularly troublesome to apply this rule to decimals, where students might think that, for example, 0.273 is greater than 0.6. This thinking actually reflects a very common misconception that students are challenged to break later (Desmet et al., 2010).	Initially, students should consistently evaluate the magnitude of a number by considering comparisons with physical materials and recording the corresponding number comparison in each case. They could use base 10 materials with whole numbers, and with decimals they could use decimal squares or a grid that they fill in to represent the area of a decimal value. In this way they can see how the digits in the number relate to the values.
The most you can have is 100% of an amount.	When learning about percentages, students are sometimes told that the most you can have of something is 100%.

This rule expires quickly in the middle grades when students engage in proportional thinking as they solve problems involving markups, discounts, commissions, profits, changes in populations, and so on. | Provide multiple examples of percentages greater than 100% to overcome any student misunderstanding. Help students see the commutative property of multiplication in action when they find that 50% of 120 results in the same answer as 120% of 50! A handy surprise for some.

Many contexts in life work with percentages greater than 100% and less than 0% (which students will eventually learn), such as financial growth, decreases in sales, and so on. |

Note: The RTEs presented in this table are synthesized from RTEs that we have collected from educators and also based on our previous work in Karp et al. (2014, 2015) and Dougherty, Bush, and Karp (2017), as well as a review by Cardone and MTBoS (2015).

We need to be clear that mathematical conventions are very different from RTEs. Mathematical conventions represent agreed-on ways of doing things that are simply "the way things are" in mathematics. You can't figure out conventions through reasoning or logical deduction; they are accepted. Some conventions vary from one country to the next. An example of a convention in some countries is that when rounding a positive number that has a 5 in the place value to the right of the place to which you are rounding, you would round up, even if the number is exactly in between the two possibilities. Another example is that some countries use a comma to separate groups of three digits to the left of the decimal point and others use a thin space. Yet another example of a convention that students encounter in elementary grades is that in a context 3 × 4 indicates three groups of four. If the problem is describing four groups of three, it would be written as 4 × 3. Interestingly, this convention is the opposite in some other countries, where 3 × 4 is interpreted as three taken four times. These are examples of conventions that help us structure our system and are assumed. So, yes, we will need to simply explain conventions such as these to students.

Now that you're familiar with many RTEs that are commonly used in elementary school, why they expire, and suggested instructional alternatives, revisit your RTE Eradication Wall. Some of the rules you listed might be included in our commonly used list, and others might be unique to your setting or experience. Use the template below to brainstorm with your team the next steps you will take to eradicate the RTEs in your school and the collective commitment you will make.

As we've been sharing throughout this book, a key component of the MWSA is ensuring that every stakeholder in your elementary school students' mathematical learning is informed and on board. While you as a mathematics teacher, instructional coach, or other instructional leader collaboratively develop the MWSA, you should also communicate the changes in mathematics instruction to parents and families as well as others involved in students' mathematical learning, such as paraprofessionals, substitute teachers, student teachers, instructors in after-school tutoring programs, volunteers, speech therapists, art teachers, music teachers, physical education teachers, STE(A)M lab teachers, and so on. Now you'll work to develop some communication tools that will inform everyone in your MWSA community to work together to eradicate the teaching of RTEs.

Mathematics RTEs MWSA sample table

Our RTEs	Our next steps for eradication

online resources Available for download at resources.corwin.com/mathpact-elementary

TRY IT OUT

Rules That Expire and Strategies We Are Using Instead in Grade _____

Rules that expire	Agreed-on instructional strategies in our Mathematics Whole School Agreement

As you let families know about the work you are doing, here is a letter that can be shared to accompany your grade-level list of RTEs. This message will help onboard parents and families or could be adapted for volunteers and other professionals in the building. The MWSA increases its potency with every person who consents to not say or promote RTEs.

THINGS TO DO

Send the Final Grade-Level Handout With This Letter

Hello Families,

We have already written to you about the Mathematics Whole School Agreement that we are developing across the entire school this year. We are now at the next phase of building our Mathematics Whole School Agreement, with this phase focused on eliminating the use of what we are calling "rules that expire." An example of one such rule that expires you might remember from when you were in elementary school is "The most you can have is 100% of something." We are no longer using this rule because we know that in real life percentages can be greater than 100% or even negative. As you can see, such a rule gets in the way of long-term learning goals. Other rules were often presented as "tricks" when you were in school, such as "Keep-Flip-Change" as a way to think about dividing fractions. All teachers and volunteers in the school are no longer using these rules that expire and are instead avoiding poorly understood tricks that either hide the meaning of the mathematics or are incorrect over time. Instead, we are teaching to increase students' depth of understanding of mathematical ideas within and across grades. Rules that expire are ineffective because they actually become a problem almost right away (expire) or as students move up the grades, and even in college or the workplace. We want your student to become an adult who will understand mathematics in whatever job they have. We are hopeful that you too can help by not accidentally saying or teaching these rules that expire. When you feel comfortable, you can help your student by using the instructional strategies we suggest. If these newer strategies are not familiar to you, please don't revert to these rules that expire! Instead, have your student explain to you how they are learning the mathematics ideas, and discuss these ideas together. We too are hopeful that you will come to sessions we are having at our school to show you the approaches we are using to teach mathematics to your student. We thank you for joining us in making this shift to eradicate rules that expire to support your student as we prepare them for their personal and professional futures!

Thank you for your help,

Your student's teachers and principal and other members of the school community

PUTTING IT ALL TOGETHER!

There are many reasons why not teaching RTEs is important. As you consider these reasons through the overall lens of the MWSA and as a leader or advocate for an MWSA in your school or district, here are some key talking points for eradicating RTEs:

- All stakeholders need to be a part of the movement that helps students overcome the idea that mathematics is a mysterious set of "tricks and shortcuts."
- Using RTEs can leave students with a collection of explicit yet arbitrary rules.
- Students tend to overgeneralize statements that teachers make to different concepts later—and they become confused and unsure.
- We want to promote conceptual understanding, and these rules often circumvent understanding a topic deeply.

CORE MWSA IDEA

When students bring RTEs to your class, be prepared and make it a teachable moment!

Now, you might be left wondering, "What do I do if a student comes to my class having already been taught an RTE?" Inevitably, even with a sound MWSA that has been firmly in place for years, this situation could still happen in isolated pockets. Even with parents and families on board, a well-meaning family member might teach a student an RTE. They could also learn one from a tutor, or a student might transfer into your school from a school that hasn't yet established an MWSA. Our advice is to make the most of it. When a student uses or presents RTEs, consider the following:

- Ask students if they can present an example that represents a counterexample to the rule.
- Ask questions to ensure that the student has conceptual understanding of the underlying mathematics especially if they are by habit using a procedure.
- Ask students to solve the problem in a different way, with different materials, and describe their thinking.
- Have another student explain why the rule doesn't work or soon falls apart.

NEXT STEPS

Continue this MWSA with us as we explore how to support students in building generalizations in Chapter 6. We will investigate how students can use the process of making generalizations to expand their ability to reason and identify patterns. Importantly, we will also consider how your instructional strategies can promote this effort by adapting tasks that can lead to significant generalizations presented in a framework to help you. We are excited for you to continue developing your MWSA with us!

BUILDING GENERALIZATIONS

Developing Instructional Strategies the MWSA Way

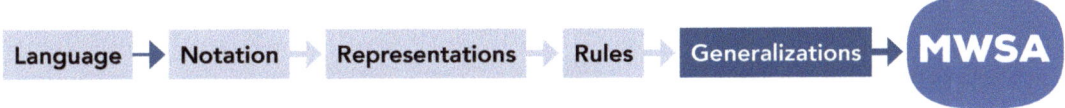

Language → Notation → Representations → Rules → Generalizations → MWSA

"All squares are rectangles."

"The product of the fraction $\frac{1}{3}$ and a whole number will be greater than $\frac{1}{3}$ but less than the whole-number factor."

"The sum of two 2-digit whole numbers will be less than 199 and greater than 19."

These statements are generalizations about important mathematical ideas, focused on the big ideas rather than on finding the solution to a specific problem. Ideally, such generalizations should be the result of students exploring, investigating, and describing the mathematical patterns they see rather than simply being told these statements by the teacher. The pathway to students formulating generalizations is guided by instructional strategies that support students' development of these big ideas.

As the next step in developing your MWSA, you will turn your attention to generalizations and the instructional strategies for helping students create them. This chapter offers guidance to support your decisions about which generalizations are appropriate for which mathematical topics and the strategies you use to develop them for your MWSA.

In this chapter you will learn

- What characterizes mathematical generalizations
- The difference between generalizations and rules

- Ways to adapt mathematical tasks to elicit students' development of generalizations

As your MWSA team works through the chapter, you are moving closer to having an established MWSA. Keep up the great work you are doing!

GENERALIZATIONS VERSUS RULES

Mathematical generalization: A claim that a mathematical property or approach is true for a large set of cases for mathematical objects or conditions.

What are mathematical generalizations? How are they different from rules? Carraher et al. (2008) define a mathematical generalization as "a claim that some property or technique holds for a large set of mathematical objects or conditions" (p. 3). Kieran (2007) described the process of making generalizations as a fundamental component of our mathematics teaching and learning. When students are developing generalizations, they look for patterns, justifying their thinking and formalizing their reasoning by representing the generalization with sketches or drawings, symbols, or words. This process is so much more than finding an answer to a problem—creating generalizations makes mathematics meaningful.

CORE MWSA IDEA

Generalizations are fundamental to mathematics!

Ellis (2011) described generalizing in three ways:

1. Identifying commonalities across multiple examples or cases
2. Extending reasoning
3. Deriving more general ideas or results from specific examples or cases

Generalizations should stem from *student* experiences and rely on meaningful mathematics tasks to help make patterns more explicit. Tasks alone are not sufficient, however; they must be combined with the appropriate instructional strategies that make constructing generalizations an organic part of children's mathematical learning experience.

When we start the process of developing generalizations with our students, we may move through at least three stages. First, students may be quick to point out that they see a pattern based on only one or two examples. Next, they may form a conjecture based on their reasoning about a relationship that they suspect is true but is not yet proven true (Lesseig, 2016). Finally, they develop a generalization that is true for multiple cases. At this point students may be able to justify the generalization by demonstrating its accuracy or validity through mathematical reasoning (Staples et al., 2012).

Conjecture: An observed pattern that is not proven to be true.

Rules (see Chapter 5), on the other hand, may not be developed through student explorations. Conversely, these may be processes that students are told to do, a "math law" they can't break. Rules may be given as a way of taking a shortcut to get students to do something quickly, though they may have little to no understanding as to why the rule works. As found in a study by Dougherty and Foegen (2014), students believe that any rule given to them by a teacher cannot be broken (e.g., PEMDAS). They believe that all such rules are truly math laws, when some are not! Rules and mathematical conventions are often confused. Mathematical conventions are used with fidelity and are consistent. Some rules, as indicated in Chapter 5, may expire as soon as they are taught.

You may have noticed in Chapter 5 that many rules traditionally used in the elementary grades are not helpful in the long run and should not be shared as math laws. These rules can be easily interpreted by students in ways that would lead them to wrong conclusions. For example, a rule such as "The longer the number, the larger the number"—often taught with whole numbers—could cause students to interpret that the rule works with decimal fractions, which we know is simply not the case. A process that was well intended by a teacher to provide structure to how students give an answer in second grade can suddenly turn into an overgeneralization that creates confusion in upper elementary grades and beyond. In this case the teacher's intention was to provide a structure for comparing numbers, not to make a rigid statement. Although well intentioned, it had unplanned consequences.

On the other hand, rules may be specifically taught with the intent that students *will* generalize them. For example, a teacher might say, "You can't take a bigger number from a smaller number" (Karp et al., 2014). As you saw in Chapter 5, this rule is not consistently true—as students will soon learn with integers when they enter middle school. Not only is this an RTE, but it is also a misunderstanding that students now have because they have overgeneralized the rule to think that they can *never* subtract a larger number from a smaller number—that it is not possible. This is another reason why it's important for those of us involved in teaching elementary mathematics to have a strong grasp of the mathematical ideas students will encounter beyond our classroom, so we can help students avoid making overgeneralizations.

There is an important difference between developing generalizations with students and giving them rules. The distinctions are subtle in some cases, and you may be wondering if you can spot the difference. Before moving on, look at the list of statements in Figure 6.1. As an MWSA team, decide which ones are rules and which ones are generalizations. You will find the answer at the bottom of the figure.

Overgeneralization: An extension or misapplication of a rule, process, or idea without justification or evidence.

FIGURE 6.1 • RULE AND GENERALIZATION SORT

A. When you divide by 10, move the decimal point over one place to the left.	B. An even number added to an even number will result in a sum that is an even number.
C. You must always start with the ones place when you are adding multidigit numbers.	D. An angle can't be greater than 180°.
E. All squares are rectangles.	F. If you are adding two 3-digit numbers, the maximum number of digits you can have in your sum is four.

Generalizations: B, E, F; Rules: A, C, D

Generalizations to Last a Lifetime

Why is having students make generalizations important? Why can't we just tell students these generalizations as rules and move on? Crafting generalizations is important because they make the big mathematical ideas explicit, ultimately helping students understand mathematics deeply and conceptually. In spite of the importance of generalizations, the actual act of formulating generalizations is even more important than the generalizations themselves. Formulating generalizations helps students connect the processes and patterns so that they have a more cohesive picture of the mathematics. The very act of formulating generalizations stems from and supports students' reasoning ability and enables them to predict or evaluate if their solution is reasonable. These characteristics are exactly what we want to develop in our students so they feel empowered by mathematics and enjoy the beauty of its many connections.

CORE MWSA IDEA

The mental action of generalizing, in essence, helps students become problem solvers, critical thinkers, and capable and confident doers of mathematics.

So what generalizations are important for students to develop? Which ones will support stronger and more robust learning? Which generalizations are important for your MWSA team to ensure that they are built and reinforced within your courses? Answering these questions is not an easy task—you may have to set aside the traditional idea that the most important thing is to develop students' procedural skills and think far beyond that. For example, the first thing that may come to mind is "I want them to learn how to divide fractions using the standard algorithm." But is that the most important idea for your students? Will that one skill support their understanding of more rigorous topics now and beyond? Enable them to reason more deeply about mathematics? To prove or justify their ideas? To become a mathematically literate

CORE MWSA IDEA

We want to develop generalizations that support learning and stick with students—ideas they can fall back on when they have long forgotten the procedural skill.

adult who can solve everyday problems? The answer is usually no; a skill can't, in isolation, improve students' mathematical reasoning ability, help them justify their thinking, or support more advanced mathematics learning. In fact, many skills "fall off" if they are not consistently practiced.

 **REFLECTION** **CONSTRUCTION ZONE—WHAT GENERALIZATIONS ARE MOST BENEFICIAL AND SPAN THE COURSES?**

Thinking about not only the grade you teach but also those before and after, consider these questions:

- Which generalizations can your MWSA team agree on that will support students' learning over multiple grades?
- Which rules are not productive in terms of helping students learn in the long run?

In the space below, record the generalizations that need to be rethought because they are currently being taught as a rule. As a starting point, you might want to consider some of the overgeneralizations made by students listed in Figure 6.1. Then, as you continue reading this chapter, other suggestions may help you reconsider what can be used in place of these.

COMMON OVERGENERALIZATIONS MADE BY ELEMENTARY SCHOOL STUDENTS

In this section, we share some common overgeneralizations made by elementary school mathematics students. Figure 6.2 provides a list of such overgeneralizations and explains how teaching these generalizations as rules may cause students to extend them to situations that do not fit with the intent of the rule. Some of these ideas were discussed in Chapter 5, but here we further expand on them to provide opportunities for your MWSA team to consider appropriate generalizations that should be included in your instruction and how they might be stated to avoid inappropriate applications. This list is neither exhaustive nor inclusive of the many overgeneralizations that may be locally based, but it should give you some food for thought as you consider others that might be found in your particular school or district. After reviewing Figure 6.2, we provide general suggestions for avoiding overgeneralizations.

FIGURE 6.2 • OVERGENERALIZATIONS COMMONLY MADE BY ELEMENTARY SCHOOL MATHEMATICS STUDENTS

Overgeneralization	Considerations for instruction
The farther away a number is from zero, the larger it is.	This overgeneralization is true for positive numbers but not for negative numbers. The overgeneralization is more related to the absolute value of a number than to the number itself (the farther away a number is from zero, the greater its absolute value). Distinguishing these nuances should be a part of refining conjectures as they move to generalizations so that students do not form overgeneralizations.
Subtraction means to decrease something because we are taking away.	This overgeneralization is similar to another one that is often developed in the elementary grades: Subtraction makes smaller. The problem with this overgeneralization is that when you are working with negative numbers, such as $-7 - (-15)$, the difference is greater than either the minuend or the subtrahend! This is an example of an overgeneralization that students may carry with them from elementary school to middle or high school.
"Improper fractions" should always be written as a mixed number.	When students are first learning about fractions, usually in the upper elementary grades, they are often taught to always change fractions that are greater than or equal to 1 to mixed numbers, perhaps so they can better visualize how many *wholes* and *parts* the number represents. While this rule can certainly help students understand that positive mixed numbers can represent a value equal to or greater than one whole, students may overgeneralize this rule and believe that it is a math law that all fractions greater than 1 must be rewritten as mixed numbers. As students soon find out as they continue to engage in algebra more formally, this simply isn't true—such as when students learn about the slope of a line and must represent the slope as the $\frac{rise}{run}$.

Overgeneralization	Considerations for instruction
A variable represents an unknown value.	Variables take on different meanings in mathematics (Küchemann, 1978), one of which is representing an unknown value. But a variable may be a specific value ($3x = 21$) or a range of values ($x + y = 23$). Establishing a variable only as an unknown quantity may cause students to act on them as if they represent one and only one specific quantity and thus not consider them as generalized quantities.

Note: The overgeneralizations presented in this table are synthesized from those that we have collected from educators as well as based on our previous work in Karp et al. (2014, 2015) and Dougherty, Bush, and Karp (2017).

Now that you've thought about some overgeneralizations that might be found in elementary school and their ramifications, revisit your generalizations list. Some of the generalizations you listed might lead to some of those included in our list, and others might be unique to your setting. Use the template that follows to brainstorm with your MWSA team the next steps you will take to instead facilitate classroom discourse in order to foster students' development of appropriate generalizations in your setting.

Overgeneralizations MWSA sample table

Overgeneralizations found in our students	Our next steps

INSTRUCTIONAL STRATEGIES FOR SUPPORTING STUDENTS IN DEVELOPING GENERALIZATIONS

As you have been working through this chapter, you have been thinking specifically about generalizations that might be helpful to students' mathematical growth and those that are not supportive. Let's turn our attention to the instructional strategies we might use to support students in developing robust generalizations.

We need to be intentional in our instruction, so that the generalizations that are developed are both mathematically accurate and applicable across different number systems or mathematical ideas. How we engage students to more closely examine mathematical ideas helps them in constructing productive conjectures and generalizations. Regardless of our intentionality, students do make generalizations (and overgeneralizations) based on what they perceive to be the common features and relationships across similar mathematical problems.

When we present problems in the same format repeatedly, students will generalize from the form they see. For example, equations at the elementary level are commonly written as $9 - 5 = 4$. When the same equation is alternatively written as $4 = 9 - 5$, students may think that this is a different problem and not recognize that they are the same relationship. In cases such as this, students have overgeneralized the equation format and may have developed tunnel vision, which prevents them from seeing these as equivalent equations.

To avoid overgeneralizations, think about how you can evaluate your instructional practice and look for these moments, as well as when and where they tend to occur. This means being particularly purposeful in the way you present each and every task.

CORE MWSA IDEA

Specific instructional strategies can lead to building strong generalizations.

Tasks to Promote Generalizations

Selecting tasks with high cognitive demand (Smith & Stein, 1998) is a critical component of lesson planning and implementation and may be part of your commitment to the MWSA. High-quality mathematical tasks can promote robust generalizations that are rich in the mathematical ideas they embody. Conversely, low-quality or ineffectively implemented tasks can fall flat in their efforts to move students beyond merely finding an answer. As you are forming your MWSA, think about the characteristics of the tasks that you would like to see presented across your grade and across your school.

The tasks we incorporate in our lessons are the pathway to students developing mathematical generalizations. Yet, often, the types of tasks we give are factual or algorithmic, as Dougherty and Foegen (2011) found in a study they conducted in one school district. Factual tasks, often in the form of a question, are those that are typically answered with a memorized fact, a definition, or a step in an algorithm, such as "What is seven times nine?" or "The name of this property is?" What may surprise you is the number of tasks or questions that can be found in a single mathematics class. In their study, Dougherty and Foegen (2011) found that in 48-minute mathematics classes in middle and high school as many as 145 factual tasks or questions were asked in a single class session.

Picture for a moment what a classroom looks like when this many factual tasks or questions are given in such a short amount of time. If you imagine a few students making choral responses or teachers answering their own questions rather than extending wait time, you are right, as that is what was often observed. However, this type of approach neither leads to substantive generalizations nor allows sufficient time and thinking room for students to explore and discuss the mathematical ideas and think deeply about the concept, all of which are necessary for the development of robust generalizations. In short, we have to move away from a focus on the quantity of the tasks or questions we present and instead focus on the quality. Before moving on, think about your MWSA team's approach to the types of tasks that are used in your mathematics classes. Do you often rely on factual tasks or questions to engage students? Do you plan the higher-level tasks and questions you are going to present before you begin teaching? Do your tasks and questions consistently focus on important mathematical ideas, concepts, and connections to prior knowledge (not just skills)?

Purposefully designing and using tasks to focus on important big ideas can lead to a much more rigorous approach to the teaching of mathematical topics as well as a way to engage your students in analyzing and thinking deeply about the discipline of mathematics. We know that your school will have an existing curriculum, but many lower-level tasks you may find there can be transformed into higher-cognition tasks to enhance your students' experience. To support you in that effort, we are proposing a framework for revising tasks that you can consider as you move forward with your MWSA.

We call this framework Process Tasks (Dougherty et al., 2015). The framework has three types of tasks: (1) reversibility, (2) flexibility, and (3) generalization tasks. These three types of tasks push students' thinking beyond just an application of an algorithm. As you go through the task types, you will notice that they require intentional design. This takes time, but the outcomes are worth it—students develop new understandings and flexible thinking. We will discuss each task type for you to think about and consider as your team builds your MWSA.

Reversibility tasks force students to think in a different direction. Sometimes, as students learn how to carry out an algorithm, a problem is given that deviates slightly from the form they have been practicing, throwing them into a tailspin. They don't know what to do, hands go up, and we as teachers may panic a little and be tempted to revert to telling them how to solve the problem. This is the point where reversibility tasks come in handy. Reversibility tasks give students only the answer and allow them to construct the task or the problem, thus developing more flexible thinking (along the lines of the TV show *Jeopardy*). For example, a traditional task might be "What is 35 + 64?" The adaptation to change it to a reversibility task would be to write, "Find two numbers whose sum is 99." Students now have new opportunities in solving the task because the adaptation opens it up to give students more entry and access points. That is, students can provide a solution based on the level and sophistication of their understanding. Some students might respond with a solution of $98\frac{2}{3} + \frac{1}{3}$, some may provide an equation similar to the traditional task, and others could give $10^2 - 1$ as a response. It is important not to interrupt students as they are thinking through a reversibility task; even if they use some strategies that are not fruitful along the way, give them time to think through the problem (Ivy et al., 2020).

Reversibility tasks: Tasks that cause students to reverse their thinking; giving the answer and having students construct the task.

There are two types of *flexibility* tasks. One type of flexibility task is to simply have students solve a problem in more than one way. The task can be left open to have students select their solution methods, or it can be structured so that specific processes are given. Student generation of multiple methods and solutions for a task provides a powerful learning opportunity as they can then compare and contrast both the solutions and the methods. However, the discussion of the solutions when more than one method is presented is to then have students compare and contrast the solution methods. How are the methods alike? How are they different? By asking students to compare and contrast the methods, they go beyond the surface of the process itself to focus on the relationships across the solution methods. An example of a traditional task is "Add 57 + 98." If we adapt it to be a flexibility task of this type, we only need to add some solution methods: Add 57 + 98. Show another way to find the sum. How are the ways you added alike? How are they different?

Flexibility tasks: Tasks that require multiple solution processes or that promote identifying relationships between and across problems.

The second type of flexibility task is to present problems that motivate students to use what they know about one problem to solve another one. The problems are structured so that the similarities between the problems are relatively transparent. If you believe, however, that students need to show all of their work or steps, you may find that this type of adaptation challenges that thinking. We think that rather than showing their work, students should be encouraged to justify their solutions and share their thinking. The purpose of this type of flexibility task is to motivate students to analyze problems before they start working on them. Let's look at a traditional task such as "Find the sum of 458 + 397." The flexibility adaptation might be as follows:

Find the sum:

1. 458 + 397
2. 463 + 397
3. 463 + 407

Traditionally, we might expect that students would add by just calculating each one as an individual problem. However, in this flexibility task, students might reason that the sums are related, as in the second addition problem one addend is 5 more than the addend in the first problem and the third problem has an addend that is 10 more than the addend in the second problem. You could ask, "How are the problems alike?" "How are they different?" "How could you use the answer from the first to find the answer to the second?" to prompt students' thinking. Tasks such as this provide insight into students' thinking and promote students' understanding of number, operation, and other relationships.

Generalization tasks: Tasks that ask students to identify and formalize a pattern from multiple cases or find a specific example given a generalization.

The third type of task, *generalization*, is specifically designed to explicitly focus students' attention on critical big ideas (or concepts). As they solve the task, students will notice some patterns emerging that they can formalize into generalization statements. These tasks may have some characteristics similar to those of reversibility tasks, but they are more explicit in asking students to make conjectures or generalizations. Another characteristic of these tasks is the opportunity to prompt students to confront common existing misconceptions related to that topic. In fact, you might use common misconceptions identified from research and prior experience as a basis for developing these tasks. Consider a traditional kindergarten or first-grade task such as 8 + 4 = ? The generalization adaptation task allows you to move beyond merely finding the sum and have students generalize about patterns with sums that lead to more significant understandings. For example, a generalization task might be as follows.

Add:

8 + 4   9 + 3
5 + 2   4 + 8
3 + 9   2 + 5

What do you notice?

You now have the opportunity to talk about the sum remaining the same when the addends are reversed. In other words, you introduced the commutative property without telling your students first! Yes, you could have told the students what to expect, but then they would not have had to engage with the same cognitive level as required by this task. Students need to be doing the thinking! The amount of thinking required to do this generalization task is higher than what is required for simplifying a given expression.

CORE MWSA IDEA

Simply applying algorithms to get an answer will not lead to generalizations.

Figure 6.3 provides more examples of traditional and process tasks to illustrate each of the three types. As you read through the tasks, stop and analyze them. What kinds of thinking do they elicit? How will the reversibility and flexibility tasks lead to students making generalizations? What generalizations will emerge from the generalization tasks?

FIGURE 6.3 • EXAMPLES OF REVERSIBILITY, FLEXIBILITY, AND GENERALIZATION TASKS

Reversibility task	Flexibility task	Generalization task
Traditional task A: 3 × 14		
Find two factors that multiply to 42.	Find the product 3 × 14 using two different methods.	What is the maximum number of digits you can expect in a product when you are multiplying a one-digit number by a two-digit number?
Traditional task B: $\frac{1}{2} = \frac{2}{?}$		
Find three fractions that are equivalent to $\frac{1}{2}$	$\frac{1}{2} = \frac{2}{4}$ $\frac{2}{4} = \frac{4}{8}$ Extend the pattern. Describe your pattern.	Siddhi said, "Two fourths is always equal to one half." Do you agree with Siddhi? Why or why not?

(continued . . .)

(continued . . .)

Reversibility task	Flexibility task	Generalization task

Traditional task C: Find the perimeter:

4 cm

2 cm

Find three rectangles that have a perimeter of 12 cm. Label their dimensions.	The perimeter of a rectangle is 12 cm. If the length increases by 2 cm, what will be the perimeter?	Cameron said, "If the perimeter of a rectangle increases, the area also increases." Do you agree with Cameron? Why or why not?

Traditional task D: Add: 13 + 14 + 15

Find three numbers whose sum is 42.	Add: 13 + 14 + 15 14 + 15 + 16 15 + 16 + 17 16 + 17 + 18 What do you notice?	Leyla said, "If I add a number to the number before it and the number after it, the sum is three times the number I started with." Do you agree with Leyla? Why or why not? Give examples to support your answer.

Traditional task E: Name this shape:

Carly drew a picture of a shape that had five faces, five vertices, and eight edges. What shape did she draw? Draw a picture of Carly's shape.	Fill in the table. Shape table below What do you notice?	Micah said, "I noticed a pattern in the number of vertices in a pyramid." What pattern might Micah have noticed?

Shape	Number of Faces	Number of Edges	Number of Vertices
Triangular pyramid			
Square pyramid			
Pentagonal pyramid			
Hexgonal pyramid			

As you read through the generalization tasks, did you find the important mathematical ideas that emerged? Check your ideas about the generalizations that can emerge for each task with the list in Figure 6.4. Use the figure to consider ways of restructuring your tasks to promote stronger mathematical understandings that lead to generalizations.

FIGURE 6.4 • GENERALIZATIONS THAT EMERGE FROM THE TASKS IN FIGURE 6.3

Task	Generalization
A. What is the largest number of digits you can expect in a product when you are multiplying a one-digit number by a two-digit number?	Students make generalizations regarding the number of digits in a product. This provides them with prediction tools to assess the reasonableness of their answers when they are multiplying.
B. Siddhi said, "Two fourths is always equal to one half." Do you agree with Siddhi? Why or why not?	Many students will say that this is true, but as they explore it further, they will notice that the relationship between $\frac{2}{4}$ and $\frac{1}{2}$ is dependent on the whole of which they are a part. The whole must be the same for the two fractions to be equivalent.
C. Cameron said, "If the perimeter of a rectangle increases, the area also increases." Do you agree with Cameron? Why or why not?	As students begin constructing or drawing a series of rectangles with a fixed perimeter, they will begin to see that there are patterns but not direct connections to calculating the area of the shape.
D. Leyla said, "If I add a number plus the number before it and the number after it, the sum is three times the number I started with." Do you agree with Leyla? Why or why not? Give examples to support your answer.	Students will notice that this is true. This task promotes number flexibility as they can think of a number as b, then the number before it is $b - 1$, and the number after it is $b + 1$. Even younger elementary students can see that the (+1) and the (−1) will result in 0, so they are left with three times the number they originally selected.
E. Micah said, "I noticed a pattern in the number of vertices in a pyramid." What pattern might Micah have noticed?	The relationship among faces, edges, and vertices is an interesting one in convex three-dimensional shapes. For pyramids specifically, the number of vertices is always one more than the number of vertices of the base.

You may also notice that the reversibility and flexibility tasks lead to generalizations in addition to the ones specifically designed to elicit them. The solutions that students create for those tasks can be used to identify patterns that can lead to conjectures and generalizations. Revising the tasks you use with the Process Task framework can dramatically change the mathematical ideas that students leave the classroom with—and the value they place on the usefulness of mathematics.

It's your turn now to start thinking about your tasks. Maybe you could begin by sharing the tasks that you use now in your classes or tasks you give that you think could be further enhanced with your MWSA team. Use the template to brainstorm with your team the next steps you will take to establish the appropriate generalizations you want to include in your instruction, and the tasks you will use to stimulate those ideas. This is also an opportunity to look at the tasks in your curriculum materials and evaluate them on their promise to elicit appropriate generalizations.

THINGS TO DO

Tasks to eliminate, with suggestions for new ones

Tasks to eliminate	New tasks to consider	Generalizations or important mathematical ideas prompted by the task

You probably have noticed that revising tasks using this framework means that your MWSA team will need to work through the mathematics content and think deeply about the mathematical ideas that you want to highlight in the tasks and the questions you might use to prompt students. Moving from using factual or algorithm-focused tasks that aim primarily at getting an answer to a set of tasks that require investigation, exploration, and deeper thinking necessitates a rethinking of how mathematics lessons unfold. Chapter 7 focuses on considerations for your mathematics lessons. The conversations your team has now regarding the changes that will be necessary to optimally use these rigorous and challenging tasks will prepare you for some of the discussions to come as you craft a meaningful MWSA.

PUTTING IT ALL TOGETHER!

You will find that this portion of the MWSA is not easy. You have to take apart the mathematics content and think beyond algorithms—this process will require time and a lot of discussion, working together as a team. These conversations may even change your own perspective about mathematics as a discipline and enhance your enjoyment of the subject! Even though it's not always simple, you will see clear improvements in student learning and the way they think about the mathematics and about doing mathematics changes. These discussions will help you discern the difference between rules (e.g., those in Chapter 5) and generalizations, focusing on students being active participants in the identification of generalizations.

Moving away from a focus on algorithmic tasks to those that develop generalizations will open the door for more engagement by your students in the "doing" of mathematics. As a result of your MWSA regarding generalizations, you'll see many benefits for your students, including the following:

- The development of more sophisticated ideas, extending beyond memorized procedures, that can support logical, quantitative, and abstract reasoning
- The ability to predict the characteristics of solutions
- Criteria by which to judge the reasonableness of answers
- Number and operation sense
- Flexibility in thinking

NEXT STEPS

As you move now to Chapter 7, you will have an opportunity to further consider how to facilitate the class discussions that will produce the generalizations needed to carry your students into higher-level mathematics and helpful career and life applications. You will see that your team's decisions in their MWSA will be helpful in thinking how you can more consistently structure your mathematics lessons. Your MWSA team is now ready for this next step!

EMBODYING THE MWSA IN EVERY LESSON

No Teaching by Telling!

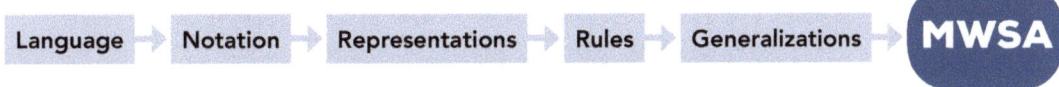

Language → Notation → Representations → Rules → Generalizations → MWSA

Imagine the following scenario. All the mathematics teachers and the mathematics instructional coach at a local elementary school are in the process of establishing an MWSA and are reading this book as part of a book study to guide their work. As they analyzed Chapters 2–6, they discussed and established some agreements and key commitments and are now very excited but maybe just a little overwhelmed about how to take the next steps. For example, they identified the less than helpful mathematical language currently used in each grade and the appropriate alternatives they will use moving forward. They agreed on some high-quality mathematical representations that they will use consistently with increasing sophistication within and across grades and have committed to using the CSA approach to structure their lesson design. With these commitments and others established, they now have questions such as the following: Do we all start the MWSA at a particular time, because I want to start some things now (or I already did)? Do we start with all components (e.g., language, notations, representations, rules, generalizations) or perhaps with just one component? What professional learning might we need to develop our content knowledge and PCK? How can we ensure that instruction is consistent but not scripted? What constitutes high-quality student learning? What should my principal expect to see if they observe any classroom embodying an MWSA?

In this chapter you will learn about

- Getting ready to implement an MWSA in your context
- Taking the different components of an MWSA you now know well and have accepted as commitments from Chapters 2–6 and ensuring that your team's instructional lessons and units embody these ideas
- Suggested appropriate high-quality and vetted resources to use in planning
- Instructional don'ts and dos for your lessons and units

In the next three chapters (7–9), you'll be engaging in the final step of establishing an MWSA.

Remember, the MWSA is about working as a team, not just as a group of teachers who show up at the same building every day. Let's get started!

GETTING READY TO IMPLEMENT AN MWSA IN INSTRUCTION

In a school that has embodied an MWSA, the mathematical success of each and every student is seen as a shared responsibility of all teachers of mathematics, instructional coaches, and administrators, and every adult who has a role in students' mathematical learning. This idea aligns with NCTM's (2014a) Professionalism principle, which states that "in an excellent mathematics program, educators hold themselves and their colleagues accountable for the mathematical success of every student and for personal and collective professional growth toward effective teaching and learning of mathematics" (p. 99). Viewing students' success as a shared responsibility and engaging in collective professional growth also align with Hattie's (2018) meta-analysis research on a construct called collective teacher efficacy. Collective teacher efficacy can be considered as the collective belief that teachers in a school share in their capability to positively affect students' learning. Importantly, Hattie found that collective teacher efficacy has one of the highest effect sizes positively correlated with student achievement.

Part of what makes an MWSA powerful is when the commitments are evident in every lesson and unit. Most often when establishing an MWSA, educators bring to the discussion their knowledge of and experiences with the existing curriculum, standards, and assessments used in their school. They might bring resources such as pacing guides or maps, detailed instructional plans, commercial curriculum materials or other vetted and nonvetted instructional resources, formative or diagnostic assessments, summative assessments or final exams, and much more. The variety of knowledge and experiences that educators bring to the conversation while establishing an MWSA may at times make the process challenging, yet it will also be extremely rewarding and absolutely worth it for educators and students alike. For this conversation to be successful, it is essential that teacher collaboration time for lesson and unit planning be nonnegotiable and there is an unwavering commitment to using appropriate and carefully vetted resources. As Iyengar (2011) stated in a TED talk, "Be choosy about choosing." Let's look at each of these aspects in more depth.

Making Teacher Lesson- and Unit-Planning Time Nonnegotiable

Leaders need to reserve a designated space and devoted time for teachers to work together as they iteratively plan and enact an MWSA. In short, "professional collaboration is critical to instructional improvement" (Larson, 2017, n.p.). This is a nonnegotiable element essential to

CORE MWSA IDEA

Teacher lesson- and unit-planning time must be protected.

success. Likewise, teachers need to leverage this time to develop and use the MWSA as a tool to inform their planning—these exchanges become a framework that informs all conversations and instructional decisions. For example, when planning a lesson or instructional unit, your team should refer to what you have agreed to in the MWSA to plan the mathematical language and notation you will use consistently across classrooms—language that will not later fall apart, expire, or undermine mathematical understanding in future grades. In another example your team might use collaborative planning time to analyze patterns in student work in order to identify potential barriers and students' strengths, which will then inform subsequent lesson plans. Here, the MWSA serves as a tool to examine why students might have used strategies that led to incorrect solutions and determine what aligned and consistent representations they should be encouraged to use moving forward. Likewise, an MWSA provides guidance on key ideas to include on an assessment, and how students might demonstrate their understanding via multiple means of expression as aligned with the Universal Design for Learning (UDL), which calls for learning environments that are productive for all learners (Center for Applied Special Technology, 2019). Teacher lesson- and unit-planning time should be used intentionally and effectively, with productive conversations. Without ongoing and ample time to have these conversations and make joint decisions, the MWSA cannot reach its full potential.

Using Appropriate and Vetted Resources

Nonvetted resources: Lessons, materials, and instructional activities that have not undergone some sort of peer review process. It is unclear if they are research informed or aligned to best practices.

As we've mentioned throughout this book, using nonvetted instructional resources, such as those on Pinterest and Teachers Pay Teachers, and those that come up simply through a Google search will likely produce a disjointed collection of teaching ideas that vary widely in terms of quality, including effectiveness and accuracy.

The danger of teachers creating their own daily lessons with instructional resources found at random through search engines is that mathematics topics are treated as isolated containers of ideas to master in a lesson or to experience through a "fun" activity. The likely result is instruction that is not deep, coherent, or aligned with a carefully crafted developmental learning sequence. The progressive nature of mathematics learning demands coherent instructional experiences that build and connect to one another, which is best accomplished through high-quality mathematics instructional materials. (NCTM, 2020, p. 39)

CORE MWSA IDEA

An essential component of establishing an MWSA is the commitment by all those involved in mathematics instruction to stop selecting nonvetted resources at random.

A report from the Thomas Fordham Institute, by Polikoff and Dean (2019), revealed important weaknesses of what they refer to as a "supplemental curriculum bazaar" (p. 1). Although the focus of this report was on high school language arts materials, the assumption is that these online sites (e.g., Teachers Pay Teachers, Share My Lessons) use the same process in evaluating materials posted in other subject matter areas as well. The expert reviewers stated that 64% of these curriculum supplements should not be used or probably are not worth using, with ratings of "mediocre" or "very poor" (p. 11). Sixty-four percent also received ratings of not aligned or weakly aligned to the standards the materials suggested they matched, and 58% either "not at all" or "weakly" built students' content knowledge and were not cognitively demanding (pp. 12, 15). In the area of addressing diverse learners, 86% offered "no supports," with another 10% offering "limited supports" (p. 16). They also found that some teachers who looked at Pinterest used the number of pins as the judgment criterion for selection. When 55% of teachers are using materials from sites such as Teachers Pay Teachers one or more times per week, these rankings are of concern. The authors drew important implications from these reviewers' findings, including "The market for supplemental materials is bewildering and begs curation" (p. 18).

The days of individually searching at 10 p.m. for fun lesson ideas for the next day's mathematics instruction are over in a school committed to an MWSA. Instead, the selection of high-quality and vetted curricular resources and materials to use in instruction becomes the collective responsibility of the mathematics teachers and other instructional leaders in the school. Instruction is planned collectively in teams far in advance and embodies the agreed-on commitments made by all teachers of mathematics in your setting. Furthermore, in a school with an established MWSA, it is acknowledged that any textbook or program adopted by the school is a curriculum resource but not *the* curriculum. Your curriculum is broader than one resource or program; it includes everything used in your mathematics instruction.

High-quality instructional ideas come from peer-reviewed and vetted sources, such as the ones listed in the next Try It Out. This list includes a sampling of books and is not intended to be an exhaustive list of high-quality resources (there are many other high-quality book and nonbook resources—e.g., peer-reviewed and vetted journal articles, curriculum materials, websites). With your MWSA team, review this go-to list, and use it as a starting place for finding vetted instructional ideas and resources to help inform your current lesson and unit planning. Decide how you want to approach this review of materials. For example, each grade level could review one resource and share with the rest of the MWSA team, or one representative from each grade can join a book review and spread the news.

TRY IT OUT

WHERE CAN MY TEAM GO FOR GREAT INSTRUCTIONAL IDEAS?

Go-to resource	Why go there?
The 5 Practices for Orchestrating Productive Mathematics Discussions (2nd ed.) Smith and Stein (2018) and *The 5 Practices in Practice: Successfully Orchestrating Mathematical Discussion in Your Elementary School Classroom* Smith et al. (2019)	These books are follow-ups to the classic Smith and Stein book *The 5 Practices for Orchestrating Productive Mathematics Discussions* (2011), which focused on anticipating, monitoring, selecting, sequencing, and connecting as ways to engage students in conversations about mathematics.

(continued . . .)

(continued . . .)

Go-to resource	Why go there?
Developing Essential Understanding Series Multiple authors—NCTM (2010–2014) and *Putting Essential Understanding Into Practice Series* Multiple authors—NCTM (2013–2019)	These are two series of books on a variety of topics for grades PK–2 or 3–5 (also middle and high school). The first focuses on the mathematics content that teachers need to know, and the second series is on how to take those ideas and put them into classroom instructional experiences. Topics include content such as number and numeration, addition and subtraction, multiplication and division, geometry, and fractions.
Developing Number Concepts Richardson (1998–2015) and *Assessing Math Concepts* Richardson (2003)	These are two series of books that support conceptual development, including diagnostic interviews and learning activities for topics such as counting, 10 frames, addition and subtraction, place value, and multiplication and division.
The Formative 5: Everyday Assessment Techniques for Every Math Classroom Fennell et al. (2017)	The authors share five proven assessment approaches with multiple examples, including observations, interviews, "show me," hinge questions, and exit tasks.
Teaching Student-Centered Mathematics: Teaching Developmentally in Grades PreK–2 Van de Walle, Lovin, et al. (2018) and *Teaching Student-Centered Mathematics: Teaching Developmentally in Grades 3–5* Van de Walle, Karp, et al. (2018)	This collection is created specifically for practicing teachers by giving information on how and what to teach for grade band topics, similar to a set of advanced-methods books. Activities, assessments, strategies, and research-informed practices are shared.
Math Fact Fluency: 60+ Games and Assessment Tools to Support Learning and Retention Bay-Williams and Kling (2019)	This is a collection of games that develop fact fluency, and corresponding assessment tools.
Becoming the Math Teacher You Wish You'd Had: Ideas and Strategies From Vibrant Classrooms Zager (2017)	This resource shares how 10 practices of mathematicians—taking risks, making mistakes, being precise, rising to a challenge, asking questions, connecting ideas, using intuition, reasoning, proving, and working together and alone—can support students' mathematics learning.
Strengths-Based Teaching and Learning in Mathematics: Five Teaching Turnarounds for Grades K–6 Kobett and Karp (2020)	This book emphasizes how identifying children's "points of power" and building instruction off their strengths rather than their perceived "gaps" can change classroom mathematics instruction for the better.

Go-to resource	Why go there?
Mathematical Mindsets: Unleashing Students' Potential Through Creative Math, Inspiring Messages and Innovative Teaching Boaler (2016)	This book provides teaching strategies and activities to support children's learning of important mathematical ideas.
Step Into STEAM, Grades K–5: Your Standards-Based Action Plan for Deepening Mathematics and Science Learning Bush and Cook (2019)	This book rightfully positions the "M" in STE(A)M at the forefront of classroom inquiries. Through multiple example inquiries, the authors empower K–5 teachers and schools to build cohesive and sustainable STE(A)M infrastructures focused on deep and integrated learning of grade-level mathematics and science.
Number Talks: Whole Number Computation, Grades K–5 Parrish (2014) and *Number Talks: Fractions, Decimals and Percentages* Parrish and Dominick (2016)	These books provide short "thinking tasks" to lead off a class session as they engage students in reasoning and strategy development.
The Mathematics Lesson-Planning Handbook, Grades K–2: Your Blueprint for Building Cohesive Lessons Kobett et al. (2018) and *The Mathematics Lesson-Planning Handbook, Grades 3–5: Your Blueprint for Building Cohesive Lessons* Miles et al. (2018)	This series of handbooks focus on the key ideas in lesson planning that work for all children.
Reimagining the Mathematics Classroom: Creating and Sustaining Productive Learning Environments Yeh et al. (2017)	The authors provide information about powerful pedagogy with an eye for equity.
Mine the Gap for Mathematical Understanding, Grades K–2: Common Holes and Misconceptions and What to Do About Them SanGiovanni (2016a) and *Mine the Gap for Mathematical Understanding, Grades 3–5: Common Holes and Misconceptions and What to Do About Them* SanGiovanni (2016b)	This book series shares common misconceptions by grade bands and how to help children navigate around them.

(continued . . .)

(continued . . .)

Go-to resource	Why go there?
Children's Mathematics: Cognitively Guided Instruction (2nd ed.) Carpenter et al. (2014)	An update of a classic book on how to support students' mathematical thinking and teaching through conceptual understanding.
Productive Math Struggle: A Six Point Action Plan for Fostering Perseverance SanGiovanni et al. (2020)	This book uses six actions—valuing, fostering, building, planning, supporting, and reflecting on struggle—to empower teachers to engage their students in productive struggle.
Access and Equity: Promoting High-Quality Mathematics in Pre-K–Grade 2 Crespo et al. (2017) *Access and Equity: Promoting High-Quality Mathematics in Grades 3–5* Crespo et al. (2018)	These books provide vignettes and ideas grounded in equity and access to empower students, engage all learners in meaningful participation, and ensure the success of each and every student.
Making Sense of Mathematics for Teaching to Inform Instructional Quality Boston et al. (2019)	This book is designed for both individuals and teams and includes a set of rubrics—the Instructional Quality Assessment toolkit—that guides reflections, conversations, feedback, and planning.
Visible Learning for Mathematics, Grades K–12: What Works Best to Optimize Student Learning Hattie et al. (2016)	This resource describes how to design impactful instruction and includes tasks and examples of instructional approaches.
Teaching Mathematics in the Visible Learning Classroom, K–2 Almarode, Fisher, Thunder, Hattie, et al. (2019) and *Teaching Mathematics in the Visible Learning Classroom, 3–5* Almarode, Fisher, Thunder, Moore, et al. (2019)	Here is a mix of strategies, tasks, and assessments that work toward students achieving deep mathematical learning in the two elementary grade bands.
Rough Draft Math: Revising to Learn. Jansen (2020)	This resource for grades 4–10 examines how students can use rough drafts to enhance their learning. By writing and talking about unfinished ideas, students are able to develop stronger understandings through multiple revisions.
Taking Action: Implementing Effective Mathematics Teaching Practices in K–Grade 5 Huinker and Bill (2017)	This book provides an in-depth discussion of NCTM's eight mathematics teaching practices, complete with vignettes, sample tasks, and classroom video.

EMBODYING YOUR TEAM'S MWSA IN INSTRUCTIONAL LESSONS AND UNITS

Let's switch gears now and dive into how your MWSA team can be sure to systematically implement and embody each component of an MWSA in all instructional lessons and units used by your MWSA team. The process of ensuring that the key components of mathematical language, notation, aligned and consistent representations, eradicating RTEs, and building generalizations are systematically implemented throughout your MWSA team is the next step toward the embodiment of a strong MWSA in your setting. Every school or community of educators is unique, so this book does not aim to be overly prescriptive about how each school should structure their MWSA process. However, in Chapter 9, we will share some helpful and practical stories from the field to inspire your continued work. Your MWSA team will need to structure the process for incorporating the MWSA in all mathematics instructional lessons and units.

An overarching goal of working together is to operationalize your MWSA team's own definition for what constitutes strong student performance. This can go a long way in helping everyone develop the same vision for your team's MWSA. While some factors will be unique to your setting, some nonnegotiables you'll need to discuss that will help frame your conversations include the following:

> **MWSA process:** The plan developed by your MWSA team to agree on, commit to, and implement your MWSA.

1. *Timeline:* What is your overall timeline for implementing your team's MWSA? Some questions to consider are as follows:
   a. Will everyone start at the same time?
   b. Will you start with specific components of the MWSA immediately?
   c. How many semesters/years will your plan take for full implementation?
   d. How will new teachers or paraprofessionals be onboarded?

2. *Format:* What template will you use for lesson plans and unit plans? Having a template does not mean that lessons are scripted. Teachers are professionals, and their personalities and the relationships they build with students are a key ingredient of effective instruction and the overall culture of your school. Consider drawing from the resources we have suggested throughout this book. Some questions to consider are as follows:
   a. How will the lesson unfold (e.g., start with a warm-up, provide time for student-centered exploration, and ensure that key ideas are discussed and summarized by the end of the class)?

    b. How will you articulate your lesson goals and objectives (we recommend mathematical learning goals as described in Huinker & Bill, 2017)?

    c. How will you introduce new mathematics content?

    d. How will you develop the mathematical practices or processes in your students?

3. *Look-fors:* Years ago in a traditional setting, a teacher was judged to be effective by having an orderly classroom with well-behaved students. Often the teacher may have been doing a lot of telling, and students may have been sitting quietly in rows of desks. But those criteria didn't always link to students' high achievement in mathematics. An MWSA classroom may look much less orderly, with students engaged in mathematical discourse in groups, moving around the room to gather resources, and sitting at tables or groups of desks to work with shared materials. Some considerations for your MWSA team are as follows:

    a. How will administrators and others know what an MWSA classroom should look like?

    b. What should someone see when they walk into a mathematics classroom influenced by a strong MWSA?

    c. What are the indicators of an effective classroom culture in an MWSA classroom?

4. *Homework:* An element of an MWSA is to craft a unified homework policy. If homework is given, we recommend focusing on the quality of problems, not the quantity, and to never give homework just for the sake of giving homework—it should be meaningful and purposeful. Avoid giving homework that requires a great deal of explanation to or by families, who are supporting the completion of the assignments. Questions to consider include the following:

    a. Do students get homework every night?

    b. What does homework consist of?

    c. How do you use homework, instructionally or for formative assessment?

    d. How does homework count toward a student's progress toward a grade?

    e. What is the purpose of your homework assignments related to student learning?

Although every school and MWSA team is unique, in the sections that follow we provide some general suggested action steps as well as ideas for teacher professional learning related to the different components of an MWSA (e.g., language and notation, representations, rules, and generalizations). These general suggestions are written for teachers of mathematics, teacher leaders and instructional coaches, administrators, and others involved in the implementation of an MWSA. These suggestions and ideas are designed to help guide and structure your continued process of embodying the MWSA in your instructional lessons and units of study. The sections below may at first appear repetitive, but these sections are written in a way that aligns to the approach we suggest you might use as you engage in this work. We suggest developing working groups made up of teachers across grades to start this process. One possible approach is for one teacher from each grade to be part of the working group on the language and notation component, one teacher from each grade to be part of the working group on the representations component, and so forth. This approach fully embodies the MWSA team mindset and empowers every teacher as a grade-level leader in a specific component of the MWSA. Representatives from the working groups for the different components can then report back and engage their grade-level teams in leading the professional learning needed to ensure that the components of the MWSA are fully embodied in daily mathematics instruction across the school. Modifications of this approach for very large schools might be to double up on representatives, and for small schools, teachers might need to engage in leading the work of more than one MWSA component.

Language and Notation

In Chapters 2 and 3 you learned about and developed commitments centered on the first steps to an MWSA, which included

- clarifying words that have different meanings mathematically in academic language from everyday conversational language,
- avoiding words that expire and using more precise mathematical vocabulary by selecting appropriate alternatives, and
- using more precise and well-understood mathematical notation, including selecting appropriate alternatives.

One of the easiest components of the MWSA to implement is refining our mathematical language and notation. This ensures that students understand the mathematical meaning of terminology that has a different meaning in the nonmathematical sense. These changes to our instruction are mostly tweaks and not overall transformations to how we teach or do mathematics. Even so, there are still changes that require intentional planning to ensure that the MWSA commitments you established in Chapters 2 and 3 systematically become a reality throughout your grade and school.

We suggest the following action steps for embodying your language and notation MWSA commitments into the structure of your instructional lesson and unit plans.

Collaborative planning

1. Take your final list of MWSA commitments for language and notation from Chapters 2 and 3, and update all existing lesson and unit plans to ensure that only agreed-on mathematical language and notation are included in any teaching materials, presentations, videos, or student consumables. Eliminate all imprecise mathematical language and notation.

2. During this process, carefully consider terminology that might be familiar to students but that has a different, and perhaps unfamiliar, mathematical meaning. Build in instructional moments to clarify such terminology. Consider instructional supports to help all learners develop more precise academic language—particularly strategies found to be successful with emergent bilingual/multilingual students and developing readers who are currently struggling.

Prior to instruction

3. Intentionally plan strategies to ensure that you use the precise language and notation encouraged by your MWSA.

4. Review the lesson plan, and consider specifically where you might inadvertently be prone to using language and notation that have been eliminated by your MWSA. Consider ways to avoid relapses. Also, intentionally plan strategies to ensure that you use the precise language and notation encouraged by your MWSA.

During instruction

5. Strive to maintain the integrity of the language and notation as prescribed by your MWSA.

6. If you accidentally use language or notation you didn't mean to, don't sweat it; something as simple as the following explanation can be used in the moment: "Wow, I didn't mean to say 'reduce the fraction.' The correct mathematical phrase that we should use is 'simplify the fraction,' because I am putting the fraction in the lowest terms." See, no big deal!

After instruction

7. Reflect on your implementation of the lesson. Did you maintain the language and notation goals of your MWSA? Consider any challenges you faced, questions you might have, and ideas for further improvement. Bring these ideas to the next collaboration time to share with others as a success story or in some cases as a cautionary tale to get ideas for new options or strategies.

8. During collaboration time, reflect on your implementation of the lesson, address challenges, consider questions, and

refine and improve the lesson plan as needed. Plan for future instruction.

How might the action steps for the language and notation components of an MWSA look in your setting?

- Third-grade teachers are sitting around a table reviewing their instructional unit on developing understanding of fractions. They are using highlighters to note the language and notation changes they need to make.
- Each grade at a school has identified words that have a different meaning mathematically from their use in informal conversation. The lead teacher then compiles them into a K–5 master spreadsheet, and the entire MWSA team develops three consistent strategies they will use across K–5 to support all learners' understanding of these words and phrases.
- To keep the process real and ensure that teachers are not too hard on themselves, the instructional coach hosts a "true confession" share-out, where team members divulge their language and notation slipups and share how they handled this in the moment during their lesson. (This whole book is our collective confessions!) The whole group discusses and identifies two go-to strategies for when this happens, as well as ways to lessen the frequency of this occurrence.
- The instructional coach and lead mathematics teacher at the school review their state's high-stakes assessment and district quarterly assessments for mathematical language, which might not be consistently introduced and used across classrooms. They share out their findings with the MWSA team, and together they identify in which lessons the terminology will be incorporate.

Representations

In Chapter 4 you learned about aligned and consistent representations to use in mathematics instruction. Replacing disconnected representations with aligned ones and ensuring consistency across topics and grades takes time and explicit effort. In the same chapter you learned how to work collaboratively in your grade and school to identify representations you will no longer use and determine which ones you will use that will be consciously aligned within and across grades, with the CSA approach at the forefront of your decision-making. Identifying your MWSA commitments for aligned and consistent representations is a big step toward ensuring that students get the mathematical learning experiences they need and deserve. With this plan in place, dedicated time will now be needed to ensure that all teachers of mathematics are comfortable implementing these agreed-on representations. This is one of the reasons why teacher collaboration time must be nonnegotiable, especially as new topics emerge as the year goes on.

Professional learning and growth are key to the success of this portion of your MWSA. Imagine this scenario. A neighboring elementary school is engaging upper elementary students in learning about categorical data and data displays. In the past the lessons on this topic primarily consisted of students answering questions about circle graphs and bar graphs from activity sheets and drawing circle graphs and bar graphs given premade data sets. This year the teachers changed their plan based on their overall focus on using the CSA approach. They started by using a percent necklace (a string of 100 beads) and student-collected data to interpret a physical circle graph. This lesson was a springboard for a class conversation about percentages, categorical data, and numerical data, and meaningful discourse around appropriate data displays for categorical data. During the next several days, students continued to engage in several data sets with rich interpretation and analysis questions to support the mathematical practices or processes, rather than working on a large number of surface-level problems. Preparation for this change in instruction called for those teaching this topic to revise their lesson plans together, gather and learn to use new materials, and talk through the parts of the lesson before implementing the change. As you can see from this scenario, designated time within the teacher collaboration sessions should be dedicated to supporting professional growth through activities such as the following:

- *Examining the agreed-on instructional representations together as a team:* This means trying out the manipulatives and instructional supports or technology, practicing drawing or sketching the agreed-on semiconcrete models, and asking and answering the questions that students may pose as they sort out their thinking. It means completing the mathematical tasks together that will be implemented with students and seeing how adaptations and differentiation can be built in, without diminishing the cognitive demand of the task.

- *Trying out the use of the newly agreed-on representation together:* Consider working in groups, setting up a mock lesson, or even watching a vetted and high-quality video modeling the strategy. The first time a teacher tests out using a brand-new representation probably should not be in front of a full class.

- *Learning through peer observations or co-teaching:* Decide that you will ask others to visit your classroom, and offer your observation skills to others. Focus on the effective implementation of the representation. A variation of this is having the coach or teacher considered the expert in that particular model co-teach with the teacher wanting to learn more. This provides support as a teacher implements new ideas. If you start onboarding new faculty with this model, it will not only support them but also change the expectations for team collaboration.

With your MWSA commitments for aligned and consistent representations established and a plan to ensure that professional learning occurs in such a way that every teacher is knowledgeable, confident, and comfortable enough to test new approaches (which might not mean completely comfortable because, let's face it, trying new things can be a little scary), the sky's the limit! Consider the following action steps for embodying this portion of your MWSA commitments into your instructional lessons and units:

Collaborative planning

1. For all existing lesson and unit plans you will be using, update them to ensure that only agreed-on representations from your commitments from Chapter 4 are included in any teaching materials, presentations, videos, or student consumables.

2. Plan for and collaboratively engage in professional learning, which includes trying out new agreed-on representations together (grounded in the CSA interwoven framework), learning through peer observation or co-teaching, planning together, reflecting on lesson implementation through examination of student work, formative and diagnostic assessments, and more.

Prior to instruction

3. Review the lesson plan, and continue practicing the use of the representation you will be using during your lesson. Consider where students might get stuck and how you will respond. Be prepared.

During instruction

4. Do the best you can to implement the agreed-on representations. If you make a mistake, simply refocus the class and try again. As stated in the book and the movie *The Martian* by Andy Weir (2014), "They say no plan survives first contact with implementation" (p. 4).

After instruction

5. Reflect on your implementation of the lesson. Did you effectively implement the agreed-on representation with fidelity? Consider any challenges you faced, triumphs you experienced, questions you might have, and ideas for further

improvement. Bring these ideas to the next collaboration time. How can these challenges or successes guide the direction of a peer observation?

6. During collaboration time, reflect on the implementation of the lesson, address challenges, identify accomplishments, consider lingering questions, and refine and improve the lesson as needed. Plan for future instruction.

 **REFLECTION**

CONSTRUCTION ZONE– INEFFECTIVE AND INCONSISTENT REPRESENTATIONS: "THE PURGE"

As an MWSA team, you've come a long way! Now it's time to release and rid yourself of the negative energy that comes with representations that you now know are characterized by the infamous "three *Is*"—ineffective, inconsistent, or just plain inaccurate. Use this wall space with your colleagues to make a list of ineffective representations you pledge will never set foot in your classroom or school again! Also list materials that linger in your classroom that should not be used and can be part of a recycling purge. Old student consumables, classroom posters, and books that focus on other structures need to go! After you are finished listing these bygone approaches, put a big, bold, powerful × through the words on this wall. Refer back to this page in the future as needed for inspiration or when helping new stakeholders learn about your MWSA. We call this process "The Purge," and it's a rite of passage that you've earned and that feels so good!

Rules That Expire

Rules used in your school that expire or simply do not promote deep mathematical understanding must exit the building! In Chapter 5 you learned about commonly used, as well as locally based, RTEs in the elementary grades; why these rules expire or fall apart; and effective instructional alternatives to use. As you know, making a collective commitment to no longer use RTEs is a substantial and essential component of an MWSA! As with the aligned and consistent representations component of an MWSA, dedicated professional learning time will be needed to ensure that every teacher is comfortable with the new instructional strategies that will be implemented in place of the obsolete RTEs, which might include following:

- *Trying out or discussing successful high-quality instructional alternatives together:* The information in the tables we provided in Chapter 5 is designed to serve as a starting point and spark your discussions. The narrative in the tables alone is, however, not always enough to prepare us to teach in a different way. Discuss these ideas with colleagues, and when applicable, try out strategies together or discuss your prior experiences that link with these ideas. Teaming up on lesson development saves time in the long run.

- *Learning through peer observations or co-teaching:* Go observe another teacher's effective implementation of appropriate instruction alternatives (instead of RTEs), or better yet, co-teach together and learn alongside each other.

As with all components of an MWSA, the commitment to not use RTEs must be consistent within a grade and across the school. It's important to remember that you can't just make a commitment not to use an RTE and expect what replaces it to be consistent, coherent, and the best choice. These decisions are not the result of individual work; they are the work of the MWSA team. For example, when it is agreed at a school that the ineffective keyword strategy will no longer be used, students will not just magically solve word problems with ease. Your team needs to work together using the effective strategies we outlined in Chapter 5 or others that you've successfully tried to build students' conceptual understanding and sense-making abilities with word problems. Consider the following action steps for incorporating the RTE MWSA commitments into your instructional lessons and units:

Collaborative planning

1. For all existing lesson and unit plans you will be using, update them to ensure that all RTE MWSA commitments from Chapter 5 are eliminated, as well as eliminated in any classroom materials, presentations, videos, and student consumables. Replace the RTEs with effective, appropriate

instructional alternatives, such as those shared in Chapter 5 and discussed during your collaborative planning time.

2. Plan for and collaboratively engage in professional learning focused on appropriate alternatives to RTEs, being prepared to implement those alternatives, and learning through observations and co-teaching.

Prior to instruction

3. Review lesson plans, and carefully prepare for lesson implementation. Consider where students might get stuck and how you will respond without reverting to using an RTE. Decide how you will handle the situation of a student bringing up a rule—what will you do? We provided some ideas for this at the end of Chapter 5.

During instruction

4. Implement the lesson. If it gets tough during the delivery of the lesson, use your plan from points 2 and 3 so that you do not revert to using an RTE. If your lesson doesn't flow as smoothly as your old favorite did, simply refocus the class and try again. Avoid giving up! And remember, your students will need time to adjust as well—stay consistent, and it will help them acclimate to the change.

After instruction

5. Reflect on your implementation of the lesson. Did you effectively implement the appropriate instructional alternative that is part of your MWSA? Consider any challenges you faced, questions you might have, and ideas for further improvement. Bring these ideas to the next collaboration time.

6. During collaboration time, reflect on the implementation of the lesson, address challenges, consider questions, and refine and improve lessons as needed. Plan for future instruction and ways to share your successes with others. For what aspect(s) would you like to have a peer observe and support you as you test new approaches?

As you continue your own personal journey as an MWSA educator, celebrate your instructional milestones! This could include celebrating the first time you teach division of fractions without using Keep-Flip-Change, when you stop using the infamous keywords as a strategy and instead focus on other strategies that develop the meaning of the problem, when you implement a new mathematical representation for the first time, or when you develop a new mathematical understanding or have an "aha" moment as a result of the MWSA journey taking place in your school! These milestones are a big deal and worth celebrating!

From RTEs to a New Way of Thinking

When seeking the best approaches to teach, again, part of the MWSA is planning collaboratively instead of immediately going "Google wild." What resources will you draw on? For example, professional organizations such as the NCTM and all its local, state, and national affiliates across North America are teams of tried and true colleagues who are on the same mission and are more than willing to share ways to respond to thorny problems such as RTEs. What follows below are some of the RTEs mentioned in Chapter 5 and some excellent NCTM articles that point to new and better approaches. Please add in your own suggestions on the supporting resources or other RTEs that need to be redirected.

Replacing RTEs with a new approach

New approaches to RTEs	Supporting articles
Equal sign	*Children's Understanding of Equality: A Foundation for Algebra* Falkner et al. (1999) *Balancing Act: The Truth Behind the Equals Sign* Mann (2004) *Using Technology to Balance Algebraic Explorations* Kurz (2013) *Early Understanding of Equality.* Leavy et al. (2013) *Fostering Relational Thinking While Negotiating the Meaning of the Equals Sign* Molina and Ambrose (2006)
Keywords	*Avoiding the Ineffective Keyword Strategy* Karp et al. (2019) *The Answer Is 20 Cookies. What Is the Question?* Barlow and Cates (2007)
The longer the number, the larger the number	*Models for Initial Decimal Ideas* Cramer et al. (2009) *Identify Fractions and Decimals on a Number Line* Shaughnessy (2011) *Decimal Fractions: An Important Point* Martinie (2014)

(continued . . .)

(continued . . .)

New approaches to RTEs	Supporting articles
PEMDAS	*A New Approach to an Old Order* Rambia (2002) *Order of Operations: The Myth and the Math* Bay-Williams and Martinie (2015)
Keep-Flip-Change	*From Whole Numbers to Invert and Multiply* Cavey and Kinzel (2014) *Measurement and Fair-Sharing Models for Dividing Fractions* Gregg and Gregg (2007) *Two Students' Constructed Strategies to Divide Fractions* Perlwitz (2004) *What Do Students Need to Learn About Division of Fractions?* Li (2008)

 Available for download at **resources.corwin.com/mathpact-elementary**

Generalizations

The final component of an MWSA is building generalizations and developing instructional strategies, which you learned about in Chapter 6. The idea of students understanding and developing their own mathematical generalizations is a part of engaging in mathematical reasoning, which includes students conjecturing, generalizing, and justifying (Lannin et al., 2011). Developing generalizations, and more broadly mathematical reasoning, is an essential part of building deep mathematical understanding, as it is how students engage in higher-order thinking.

For MWSA teams to teach in ways that focus on students' ability to build generalizations, they must pay heed to intentionally building all teachers' deep mathematical and professional growth through the following:

- *Developing teacher content knowledge and PCK:* The importance of planning and working on this part of the MWSA as a team cannot be overstated. Teachers should practice tasks and problems together, discuss the mathematics content, and rehearse the asking of higher-level questions together.
- *Trying out new instructional strategies together:* Teachers might work in groups, setting up a mock lesson or even watching and discussing a vetted and high-quality video that models the strategy.
- *Learning through peer observations or co-teaching:* Decide that you will ask others to visit your classroom, and offer your observation skills to others. Focus on the effective implementation of an instructional aligned strategy, and note the effects on student learning, especially engagement, discourse, and evidence of student understanding. A variation of this is having the coach or teacher considered the expert in that particular strategy co-teach with the teacher wanting to learn more. This model provides support as a teacher implements new ideas.

With your MWSA commitments in place for building generalizations and developing instructional strategies and a plan developed to ensure that professional learning occurs, consider the following action steps for embodying these new commitments in your instructional lessons and units:

Collaborative planning

1. For all existing lesson and unit plans you will be using, update them to ensure that only agreed-on MWSA commitments from Chapter 6 are included in any teaching materials, presentations, videos, or student consumables.

2. Plan for and collaboratively engage in professional learning, which includes trying out new agreed-on strategies, learning through peer observation or co-teaching, planning together, reflecting on lesson implementation through examination of student work, formative and diagnostic assessments, and more.

Prior to instruction

3. Review lesson plans, and continue deciding on the instructional strategies you will be using during your lesson. Consider where students might get stuck and how you will respond. Be prepared also for students' success.

4. Design reversibility, flexibility, and generalization tasks by adapting some of your traditional tasks via the Process Framework.

During instruction

5. Do the best you can to implement the agreed-on instructional strategies. If you feel tentative, simply refocus the class and try again. This part of the MWSA will take time, and it may not go perfectly at first—or then again it may!

After instruction

6. Reflect on your implementation of the lesson. Did you effectively implement the agreed-on instructional strategies? What did you notice about your students' learning and depth of mathematical understanding? Were they forming new connections? Consider any challenges you faced, questions you might have, and ideas for further discussion with colleagues. Bring these ideas to the next collaboration time. How can these challenges guide the direction of a peer observation?

7. During collaboration time, reflect on the implementation of the lesson, address challenges, consider questions, celebrate successes, and refine and improve the lesson as needed. Plan for future instruction.

How might the action steps for the generalization component of an MWSA look in your setting?

- Primary grade teachers work together to change the cognitive demand of their tasks. For the next day's lesson on addition, they all agree to use the two reversibility tasks and one generalization task that they had developed together by adapting their existing curriculum.

- At the end of each month, the grade band teams at a local elementary school have a friendly competition. Each team presents to the group their best reversibility, best flexibility, and best generalization tasks—with discussion of student thinking and work samples that were elicited from those tasks. Everyone in the entire MWSA team votes, and winners get bragging rights. Most important, the whole team is learning and continually improving the tasks that are adapted and developed.

- At the elementary school across town, Friday MWSA team time is dedicated to not only sharing out key samples of student work but also highlighting incorrect and correct solution strategies. This consistent practice builds teachers' collective skills to analyze student work, provides examples to use in class, and builds generalizations to support the next instructional steps.

PUTTING IT ALL TOGETHER!

This chapter was about the process of embodying the MWSA commitments your collaborative team made in Chapters 2–6 in instructional lessons and units. In this chapter you learned considerations, steps, and strategies for taking your MWSA from an agreed-on team commitment to the embodiment of each component in your instructional lessons and units. Now it's time to pause and check Figure 7.1 for a summary table of don'ts and dos as you work to exemplify your MWSA in your mathematics instruction.

FIGURE 7.1 • INSTRUCTIONAL LESSONS AND UNITS: DON'TS AND DOS TO REMEMBER

Instructional lesson and unit don'ts	Instructional lesson and unit dos
Don't think that if you taught it, it was learned.	Focus on evidence of student learning.
Stop mini lessons, which may be good for reading but are not successful in mathematics.	Build a coherent sequence of lessons.
Don't think that students must complete a set number of problems or an entire worksheet.	Focus on the quality of mathematics tasks and instruction, and the use of multiple strategies on the same problem, not on the quantity of problems.
Avoid teacher-centered instruction, such as I do, we do, you do (see Berry, 2018), or gradual release of responsibility (see McCaffrey, 2016).	Employ student-centered instruction that emphasizes you (students start thinking about the problem on their own), y'all (students share with a partner or at a table), and we (the class discusses the ideas and strategies that emerged) (Lampert, as cited in Green, 2014).

(continued . . .)

Instructional lesson and unit don'ts	Instructional lesson and unit dos
Don't require teachers to follow an exact script when teaching a lesson.	Adhere to your MWSA, and use curricular materials with consistency and fidelity
Eliminate inconsistent formats for lesson and unit plans across the team.	Develop a consistent format for lesson and unit plans across the team.
Don't focus on only one way to solve a problem to reach one right answer.	Focus on students using multiple solution paths and strategies, showing more than one way to approach a problem and in some cases identifying multiple correct solutions.
Don't focus only on correct answers.	Focus on explaining, justifying, and supporting reasoning.
Don't replace mathematical language and definitions with cute sayings, acronyms, and mnemonics.	Incorporate precise and agreed-on mathematical language and definitions into instructional units, and have students use this terminology.
Don't teach students tricks, shortcuts, and rules as a replacement for deep conceptual understanding or procedural fluency.	Develop students' procedural fluency built on the foundation of conceptual understanding and understanding the meaning of all procedures.
Avoid using only abstract symbols.	Use physical or concrete materials, semiconcrete representations, and abstract symbols to model mathematical ideas.
Don't sequence the representations so that students see only one at a time.	Use the CSA representations concurrently.
Don't use materials or illustrations without carefully assessing the affordances and the potential for misconceptions.	Select the representation that most accurately represents the mathematical idea.
Don't select disjointed resources from nonvetted outlets.	Use vetted, high-quality, and coherent instructional resources.
Avoid planning instruction in ways that privilege a few students and marginalize many.	Establish a commitment to high-quality and equitable instruction and to adopting all of NCTM's (2014a) eight mathematics teaching practices.
Don't teach in ways that primarily focus on memorization, speed, and procedures, without understanding how the procedure works.	Embody a commitment to developing deep mathematical understanding of all operations and procedures.
Don't attempt to "water down" instruction or not teach on grade level as a way to help students.	Establish a commitment to teaching each and every student grade-level content, with the instructional support and differentiation they need to be successful.
Don't provide scaffolds for students that might diminish the cognitive demand of a task, referred to as "just-in-case" scaffolding (Dixon, 2018; Dixon et al., 2019).	Provide scaffolds for students that maintain the cognitive demand of a task, referred to as "just-in-time" scaffolding (Dixon, 2018; Dixon et al., 2019).
Avoid planning independently.	Engage in collaborative planning and team efforts.
Avoid staying deeply rooted in traditional teaching methods that are disconnected from what today's mathematically literate members of a democratic society need.	Plan lessons that are informed by reform-based and research-informed best practices in mathematics education to support students long beyond their PK–12 experience.

Instructional lesson and unit don'ts	Instructional lesson and unit dos
Don't jump straight to the algorithm without the underpinnings that provide understanding.	Introduce concepts before procedures, delaying algorithms.
Don't reveal that mathematics isn't your favorite subject or that you didn't do well in mathematics as a child.	Share a passion for mathematics.
Don't focus on how quickly a student can get an answer.	Allow adequate wait time.
Don't demonstrate how to use manipulatives and have students copy the presentation exactly.	Embed the use of manipulatives and concrete materials in students' problem-solving lessons.
Don't present tasks focused on algorithms or skill or broken into small parts.	Give tasks with higher cognitive demand using reversibility, flexibility, and generalization (Dougherty, 2001), as described in Chapter 6.

NEXT STEPS

Continue this MWSA journey with us as we explore enacting the MWSA in Chapter 8 and expanding it beyond your school to other schools in your district and other stakeholders. This next step is crucial in the MWSA becoming the norm and culture in mathematics education. Join us as we collaboratively continue this journey!

GETTING TO THE HEART OF THE MATTER

Building and Enacting the MWSA

Two teachers are sitting in the teacher workroom talking about their mathematics classes. One teacher says, "I had a student tell me today that 2 + 5 and 5 + 2 are turnaround facts. Have you ever heard that before?" "Yes," said the other teacher. "I've heard the commutative property called all kinds of things. My students act confused when I introduce the more accurate mathematics vocabulary." Do conversations such as this one sound familiar? Have you, just as these teachers, noticed that students come into your class with mathematical strategies, such as factor rainbows, that have a less than effective impact on their learning? Do students have a difficult time acclimating to your mathematics class or lesson structure? If you answered yes to any of these questions, an MWSA will inspire and lead to positive change in your mathematics program schoolwide, and even districtwide!

In this chapter you will learn

- Important factors to consider in the change process, sparked by your desire to start an MWSA
- The initial steps you can take to enact the MWSA process

Bringing your decisions and commitments all together is a big step for your MWSA team. The results of your efforts will be worth it when you see the effect it has on student learning and engagement. Get ready, get set, go!

ENACTING CHANGE

Enacting your MWSA requires a few basic building blocks—most important, thinking about strengthening collaboration and navigating change. Before we get into the heart of the matter, let's look at how to successfully engage in collaboration and change.

Collaborative Teamwork

Collaboration is a process that leads toward a shared goal. It bridges isolation and strengthens our ability to try out the new ideas that a colleague has attempted and succeeded in doing. In fact, it is a necessity in enacting the MWSA and can multiply the powers of the individuals involved in exciting ways. Any effort that involves working as a team entails trying out new approaches, and when an attempt isn't successful, it entails revising and trying again.

You are likely approaching this work as an administrator, an instructional coach, or a teacher. Here is what everyone on the team needs to do in order to set up a strong collaboration as your school or district prepares to embark on enacting your collective MWSA:

- *Be planful:* Set regular meetings with clear agendas and specific outcomes.
- *Be inclusive:* Ensure that everyone has opportunities to reflect and contribute—possibly first as individuals or pairs, then in small groups, and finally as a large group.
- *Be open:* Commit to being open to change and ready to try out new ideas and approaches toward building cohesive delivery of mathematical content, practices, and processes.
- *Be encouraging:* Celebrate milestones as you move toward shared objectives.
- *Be flexible:* Use compromise as a tool to bridge differences in teaching styles and philosophies.
- *Be accountable:* Give and take fresh ideas, try them out, and circle back with your colleagues.
- *Be responsible:* Think of your colleagues' students as your students—share responsibility for all learners.
- *Be focused:* Always center the conversation on what is best for students, not just for now but for their future.

Planning for Change as a Team

Mathematics curriculum and pedagogy may seem ever-changing, causing novice teachers to report feeling overwhelmed and seasoned teachers to say, "I was here before *it* got here, and I will be here after *it* is gone!" Change is hard, especially when the change is ingrained in what many teachers believe to be the "way math is taught." It's hard to refute 13+ years of classroom instruction that reinforces what seems to be a logical model of practice. But the standards have changed, students' needs have changed, the jobs that exist today have changed, and the world has changed! So what do we need to know from research about the process of effecting and enacting change?

If we want to understand how to encourage change not only at the individual level but, more important, also at the team, building, and district levels, let's look at some background on approaches to managing change, resolving conflict, and collectively solving problems. Lewin (1947), who was grouped with Freud as being one of the most influential researchers in psychology, coined the term *group dynamics*, which explored the significant value that group experiences provide in directing members' actions and decisions. He suggested that groups are the main focus of change, not individuals. Therefore, let's look at an approach based on Lewin's three-step framework for changing group behaviors (Rosenbaum et al., 2018):

1. *Unfreeze the mind:* Although the mind is never really *frozen* and is continuously in action, the school team stirs up a critical eye on previous approaches to create disequilibrium. This stimulation provides the edge that invites the emergence of a willingness to change.

2. *Moving and changing:* The group jettisons ill-functioning approaches for a more beneficial and sound collection of behaviors by building consensus and growth through iterations.

3. *Refreezing the group in a new location:* The team stabilizes the new set of reconfigured behaviors, resulting in a new array of stronger practices and norms. All agree, "We don't go back (unless we have new data that suggest that we should)!"

Although this pattern exists, the overall model is not linear in its application or its success as different pieces and steps in the process need to be iterated, revisited, and reexamined. Change in schools can sometimes be very rapid and other times incremental and almost imperceptible, like the movement of a glacier.

As stated by the former United Nations ambassador Samantha Power, "Sweeping change actually usually comes as a result of incremental changes" (as quoted in Luscombe, 2019, pp. 22–23). So what should happen schoolwide—not just in your grade? Begin by completing the next Things To Do in advance, and then set up a gallery walk (with multiple copies of each of these three components on chart paper) around the room, where teachers, instructional coaches, and administrators can share their thoughts with the group.

THINGS TO DO

Planning for the Three-Step Change Framework

Things to stir up	Easy targets for things to jettison	New practices and norms you'd like to see

online resources Available for download at **resources.corwin.com/mathpact-elementary**

Part of building cohesiveness across the school is celebrating the good work on the MWSA that people do. Those people can be your colleagues, but sometimes it may be just you! Here is a way to formally acknowledge your milestones along the road of working on a team at your grade level or across the school. Note that we not only focus on noticing these indicators of success, but we also ask that you think about how they can be officially recognized. We also know that sometimes you are making changes on your own that push you out of a comfort zone or represent the altering of long-held practices that were fixtures of your teaching. We respect these hard efforts, and we want you to rightfully celebrate these landmarks with all sorts of positive kudos and commemorations.

THINGS TO DO

Celebrating Personal MWSA Milestones

Use this space to keep track of your personal MWSA milestones and document how you will celebrate! Celebrating these signposts of your victories can be done in simple ways such as sharing your success with a colleague, taking a walk, or buying a new book, or perhaps your MWSA team might even come up with a creative way to commemorate personal high points! Keeping a record of these personal milestones is important because it reminds you how far you've come, that the hard work you are putting into the alignment and change process is making a difference, and it helps you recognize the strengths you bring to the collective group (for more on a strengths-based approach that acknowledges your strong points, check out Kobett & Karp, 2020).

My personal MWSA milestones

MWSA milestone	How I will celebrate

 Available for download at resources.corwin.com/mathpact-elementary

Why do people resist change? For many of us, it's hard to make change when traditions are so deeply rooted that we are on automatic pilot when we talk and teach. For example, we suggest that all MWSAs should include an agreement to not use the phrase "reducing fractions," which we discussed in Chapter 2. For people so completely accustomed to using this phrase, this may be a hard shift to accept and make. After all, we've said "reducing fractions" so often and heard it so many times in our life, it rolls off our tongue like a spontaneous reflex! But what happens when we decide that this expression doesn't make sense because there is no real reduction of fractions—they are not getting smaller or going on a diet? That's when we all sit back and reflect and say, "Of course, that doesn't make sense" and then "Why was I taught that?" Through such reflection we become more open to adopting a more carefully thought-out mathematical communication or phrase such as "rewriting fractions in the lowest terms" or "simplifying fractions."

We need to assess readiness for change through a more mindful approach (Gondo et al., 2013). Mindfulness is paying close attention and developing an awareness of what is taking place right now (Brown & Ryan, 2003). Gondo et al. (2013) found that we hold "enduring assumptions" and beliefs about "how we do things around here" (p. 37) that stick around long after the meaning they initially supported has changed or is no longer appropriate. If we can mindfully examine our everyday decisions and actions, we can embrace the change process and assess how to alter our behaviors. An MWSA is not a makeover—it is a make better! Let's look at the following reflection to see how a shift from exploring explicit factors for change to examining how to alter thinking based on well-worn patterns, norms, and routines might play out. We start here with a few examples (see Figure 8.1), and you can add others that ring true for your personal experience or the experiences you have supporting your colleagues.

CORE MWSA IDEA

An MWSA is not a makeover—it is a make better!

You may find that some colleagues are still resistant to change. This is natural, so what can you do? Consider what information or data these colleagues might find compelling. As you will see in Chapter 9, there are many approaches to respond to this dilemma. Some find a way to collect data from students that provide the impetus; others go as a team to attend a conference and hear ideas they want to try. As more and more of your colleagues join in, the sheer presence of a strong and positive peer group may also be a motivating factor.

FIGURE 8.1 • RESISTING AND EMBRACING CHANGE

Reasons why we might resist change	Benefits of embracing change
I don't have the time to replan my lessons.	An MWSA isn't a solo activity. By working as a team, we share in the responsibility, the work, and the benefits. In the long run, the MWSA way will save us all time!
I have too much on my plate already; I can't add one more thing.	While our MWSA will take some work up front, in the long run it will take so much off our plates. It will remove time spent reteaching, undoing, and fixing instruction.
I learned mathematics procedurally and did just fine, so my students can too.	While this is true for some, it isn't true for all. Not all students are just like us. What are the richest and most equitable learning experiences we can offer our students, using the ideas in Chapters 2–6? What do we owe our students?
I've always taught this way.	This goes back to when we learned that if we know there is a better way we need to work toward it. That's our professional responsibility. We can't say that we are lifelong learners if we don't step forward to make changes that will enhance our students' understanding.

As you and your team have learned throughout this book, enacting an MWSA will require many discussions and subsequent agreements. The process you use in creating your MWSA has to allow for critical conversations. Be prepared—change is difficult! But the time you invest now will pay dividends in the future.

MOVING MWSA BEYOND YOUR SCHOOL

Now that you've learned about the MWSA and are enacting it in your school, you are ready to embark on the final step in the MWSA process—moving the MWSA beyond your school. Establishing an MWSA across a school district and in the community involves broader groups in adopting an entirely different mindset about what mathematics instruction looks like, what it means to learn and know mathematics, and the role families and students themselves play. For example, traditionally students' mathematics homework can be a point of anxiety and stress for both students and families alike. Parents often feel that they cannot help their students do "the type of mathematics they are doing today in schools." We've seen on social media many posts and memes that state that mathematics is "different from what it was when we were in school" or that it's "much more complicated than it has to be." Traditionally, people jump to what they feel is comfortable, and it is no surprise that parents and families jump to algorithms and understandably may get frustrated with schools for using mathematics strategies that are unfamiliar. In the past, success in mathematics has often been reserved for students perceived as being "good at math." More broadly, it has been culturally acceptable to say "I can't do math," "I was never good at math when I was in school," or "Math was never my strong subject." All this negative talk must *stop*! While this avoidance of "dissing" mathematics is critically important at all grade bands, young children are at such an impressionable time in their lives when they are beginning to shape who they will become as adults, so we must be especially careful in the elementary grades. Let's do all we can to shape them into confident and capable mathematicians!

Now imagine a scenario, one where it's clear why students are learning the mathematics they are learning. Aligned representations and consistent instructional strategies used in classes are transparent not only to teachers but to all stakeholders involved in students' mathematics education, including parents and families and school community members. Precise and appropriate mathematical language and notation are used, and tricks, poorly understood shortcuts, and RTEs are simply not part of an effective mathematics instructional program. Students learn mathematics for the ultimate goal of becoming mathematically literate, STEM-savvy, and well-informed members of society with a passion for mathematical ideas. Everyone knows the MWSA and is committed to its success. Mathematics is approached collaboratively as a community endeavor. Just as with a school sport, academic club, or band, the entire community—both within schools and beyond—rally together and are excited to support the home team, celebrating each and every student's mathematical growth and success. The support is unwavering, and all stakeholders know that the investment, although great, is well worth it for students' current and future personal and professional lives, as well as for the good of the community at large. This amazing scenario is one you are in the process of creating as your team enacts your MWSA! We are sure that you are already seeing this within your school and in some conversations with parents and families.

In this section, we shift our attention away from thinking of MWSA at the school level and begin to explore how to (1) establish a horizontally articulated MWSA across the schools in a district (i.e., consider how children move from preschools and day care to all elementary schools in the same district), (2) launch a vertically articulated MWSA across the schools in a district (in this case, calibrate as students transition from elementary to middle and from middle to high), and (3) ensure that all stakeholders are involved.

CORE MWSA IDEA

An MWSA helps everyone to be on the same page—a win-win!

Establishing a Horizontally Articulated MWSA

All students need and deserve access to the highest-quality mathematics education. This access cuts across all teachers and classrooms within schools as well as across schools in a district. For example, all elementary schools in a district should teach students about fraction concepts, using the same high-quality and conceptually based models and strategies. Just as labels such as "low–math ability student" and "high–math ability student" are completely unacceptable, so are labels such as "the best elementary school" or "an elementary school you don't want to send your child to." Families and students should believe, and it should be true, that every school in the district offers the best mathematics instruction. This is why it is important for an MWSA to exist not only within one elementary school but also across all elementary schools in the district. This process is simply doing what is right for all students, especially for those with high mobility within a district, as they will arrive at a new school already knowing the mathematical language, notation, representations, generalizations, common instructional strategies, and norms. What a wonderful thing it would be for a student to feel "at home" in mathematics class when other aspects in their new school are unfamiliar. While there is no single prescribed way to go about establishing an MWSA across all elementary schools, consider the following helpful hints and suggestions:

- *Coordinate the collaboration process:* District-level mathematics/curriculum leaders should coordinate the process of establishing a districtwide MWSA in collaboration with building administration and a designated leader in each building. If no school already has an established MWSA (all starting at the same time), get every school involved from the beginning. If one or more schools already have an established MWSA, work to build from those ideas, and use consensus building and make choices based on research-informed best practices.

- *Support from building administration:* Building administration should be completely on board, be trusting of the process, and support it through providing collaborative time to establish (across and within schools) and implement the MWSA. Administration also should understand the tenets of the MWSA and recognize and support MWSA instruction during informal and formal observations and evaluations.

- *Establish a building leader:* Designate a leader within each building to spearhead the effort and coordinate the process. This person could be a mathematics instructional coach, department chair, or grade-level lead teacher.

- *Create a timeline:* Don't rush the process, but set firm goals and work hard to meet them. This structure helps ensure that things keep moving forward in the process. Establish the agreement in the same order that we've recommended throughout the book—language and notation first, representations second, eradicating RTEs third, and finally working on generalizations.

Moving the Process Across Schools

Consider how you might advocate for or start the process of establishing an MWSA across all elementary schools. If you are a principal, you might start with a conversation with other principals or the district mathematics specialist. If you are an instructional coach, you might ask your principal if you can go together to talk to the district administration. If you are a classroom teacher, you might talk to an instructional coach or an administrator in your building. You know your school and district structure best, so wherever the conversation starts, the most important thing is that you begin. After you get the conversation going, we suggest adopting a systemic approach so that your district's MWSA work is transparent, collaborative, and inclusive. The previous list provides good beginning guidance for putting structures in place. In this space jot down three ideas you have for getting the conversation or process started for establishing an MWSA across all elementary schools.

1.

2.

3.

Establishing a Vertically Articulated MWSA

Another part of the broader elementary school MWSA process is ensuring consistency and coherence vertically as students transition from preschool and day care to elementary school, then on to middle school, and finally to high school. Accomplishing this is part of the districtwide MWSA, which we admit can be complex. Students should begin learning mathematics grounded in the ideas of an MWSA even before arriving at kindergarten. For those of you who have 4-year programs within your district's jurisdiction, you are ahead of the process of building connections. Others among you may know the local pool of preschools and day care centers and make explicit invitations for those leaders and teachers to join you in an evening discussion of your MWSA. This preliminary work is important and will support your incoming students in the ways you want these children to join the school, aligned with the best possible groundwork for learning mathematics.

With a strong foundation of mathematics understanding built in the elementary grades through a strong elementary mathematics MWSA, students will be ideally positioned and ready for the increasingly abstract and algebraic mathematics of the middle and high school courses. Elementary, middle, and high schools often do not communicate as much as they would like to about their mathematics programs. The importance of a coherent mathematics program for PK–12 and beyond cannot be overstated and aligns with the ideals of *Catalyzing Change in Early Childhood and Elementary School Mathematics: Initiating Critical Conversations* (NCTM, 2020). In other words, the MWSA established in elementary schools should align (with regard to language, notation, representations, agreement not to use RTEs, generalizations, strategies, and so on) with their feeder middle schools, and the MWSA in middle schools should align with their feeder high schools. While the mathematics becomes more complex and sophisticated, the ways in which students learn mathematics should be consistent, familiar, and cumulative. For example, in the elementary grades, students should use an area model as a way to build understanding of the distributive property when multiplying multidigit whole numbers. Later in elementary school and early in the middle grades, students can use this familiar area model to build understanding of the distributive property when multiplying fractions. Soon after in the middle grades and into high school, students use the area model once again to consider algebraic applications with the distributive property when multiplying binomials or trinomials. This strategically aligned model builds coherence as the same idea increases in complexity while maintaining familiarity with an area model.

Take a minute to think about some of the ways in which mathematics should be coherent across PK–12, such as the idea we shared with the area model example. The ideas you consider could be consistent strategies to use across grade bands or aligned models, or for building ideas coherently as students move up through the grades. This is your time to brainstorm, perhaps with a partner or your team. Record your ideas in the reflection box!

In Figure 8.2, we provide some ideas to help you think about the commitments that might be part of a PK–12 MWSA. Compare these with the ones you discussed in your collaborative team, and then continue your consideration of other commitments.

FIGURE 8.2 • SAMPLE COMMITMENTS IN AN MWSA ACROSS PK–12

Commitment
Strategies for teaching word problems (avoiding a keyword approach) are aligned across the grades.
Algorithms are delayed until conceptual understanding is reached.
When properties or other mathematical ideas are introduced, the precise mathematics vocabulary is used to name or describe them.
Students have opportunities in every lesson to describe and explain their thinking and reasoning.
Physical or concrete, semiconcrete, and abstract representations are presented concurrently.
More than one algorithm is taught to provide more options for students as part of their sense-making.
Students are often asked to show their thinking using multiple representations or in multiple ways.
A foundational commitment to teaching through problem-solving and sense-making is prioritized.
Higher–cognitive demand tasks and questions will be commonly used in instruction.
Students develop generalizations organically rather than being taught them by the teacher.
Rather than give students 20 problems to solve, give 5 problems that have to be solved in two different ways.
Use a carefully defined trajectory of representations so that at each grade level or in secondary school teachers know what came before and what comes next for each mathematical domain.
Ensure that symbols such as the equal sign, for example, are well understood.
Make mathematics memorable so that students learn to love this essential subject.

These commitments will drive your instructional decisions and ensure that all students have equitable access to understanding mathematics. Revisit this list as you move forward in broadening the MWSA.

ENSURING THAT ALL STAKEHOLDERS ARE INVOLVED

When thinking about an MWSA, it's easy to see where teachers of mathematics, instructional coaches, and school and district administrators fit into the conversation. While we've mentioned parents and families in our previous chapters, we want to be clear that all stakeholders involved in students' mathematics education should be well informed so that they can engage in the process and continue to support students. Such stakeholders include the following:

- Teachers of mathematics at any grade level
- Special education teachers
- Other teachers involved in teaching mathematics, such as teachers of emergent bilinguals
- Instructional coaches
- School and district administrators
- School academic counselors
- Paraprofessionals
- Teaching assistants
- Other teachers and staff in the building, such as the physical education teacher, art teacher, music teacher, foreign language teachers, speech therapist, and occupational therapist
- Student teachers/preservice teachers
- Long-term substitute teachers
- Interventionists
- STE(A)M lab teachers
- School volunteers
- Parents, grandparents, guardians, and family members
- Instructors in before- or after-school tutoring or club programs

The engagement of this broad group of stakeholders will ensure that students are supported in all of the mathematics work they do. You also need to consider how you will bring on board new faculty and administrators. Their perspectives may also help you think about aspects of the MWSA that were not considered in your initial discussions. Consider asking those you are interviewing for positions about their willingness to participate in this cross-school (or district) approach.

PUTTING IT ALL TOGETHER!

 **REFLECTION** **CONSTRUCTION ZONE– PROFESSIONAL GROWTH**

This chapter has focused on operationalizing the MWSA process within your school and then broadening the commitments across the schools in the district and beyond. Stop now, and reflect on the professional growth of your grade-level team, school, or district (depending on your role). Record your thoughts by finishing the following prompts.

When reflecting on our mathematics MWSA journey:

We are most proud of . . .

_____

_____

_____

We are most surprised by . . .

_____

_____

_____

We still need to work on . . .

_____

_____

_____

NEXT STEPS

We hope that this process has been professionally engaging, filled with new ideas and understandings. In Chapter 9 you will be inspired as you read others' stories of their MWSA process and outcomes. As you compare your road map or your journey with theirs, think about the exciting opportunities you are giving students to become confident and competent mathematics students—and what a difference you are making!

SHARING SUCCESSES FROM THE FIELD

MWSA Heroes Tell Their Stories

Take a moment to reflect on your team's MWSA progress. We imagine that now, as a result of your hard work and perseverance, the mathematics program at your school is transforming in wonderful ways. For example, students and educators are now using (or have plans to use) appropriate mathematical language and notation; there is coherence, consistency, and depth in learning mathematics with aligned representations; RTEs are in the rear view mirror (thank goodness!); there is a new commitment to building students' ability to make generalizations and reason effectively; and lesson structure is aligned to the CSA framework. As your school collectively continues this important work, your MWSA will be further refined and should be considered a document that will be forever evolving. As you read earlier in this book, once we know better, we have a professional obligation to do better—and just think about how much we have all learned!

This final chapter focuses on educators who have implemented all or some of the components of an MWSA. They are the early adopters and the heroes who started the idea in their setting with one or two people. From that beginning their efforts expanded to a full MWSA team that created strong mathematics learning experiences across the grades.

In this chapter you will learn about MWSAs in action through multiple settings:

- How sending a group of teachers to a conference led to a schoolwide transformation
- How one educator spearheaded an MWSA movement in their school
- How the guidance and vision of a single mathematics coach led to a remarkable schoolwide transformation

Remember, the MWSA is both a destination and a long-term team journey. Let's get started on the penultimate leg of this trip!

EXAMPLES OF MWSAS IN ACTION

The MWSA, both within and across the schools in a district, is a powerful tool for rich and transformative conversations during teacher collaboration time. Making decisions about what will be included in an MWSA and moving it to implementation serve as a catalyst for conversations that may not otherwise happen. In this section we provide three detailed stories of MWSAs in action to make the process more concrete, to let you know that you are not alone, and to provide inspiration! Remember that every setting is different, so while these stories are great examples, they are not designed to be prescriptive.

CORE MWSA IDEA

An MWSA is both a destination and a long-term journey.

A Conference Session Leads to Teachers' Action: A Small Group of Thoughtful, Committed People

Never doubt that a small group of thoughtful, committed people can change the world. Indeed, it is the only thing that ever has.

—Margaret Mead, cultural anthropologist

This group of thoughtful and committed people work at Discovery Elementary School in Northern Virginia. We now share their story. The mathematics coach, Angela Torpy, and the elementary principal, Erin Russo, made up the core group that made the MWSA happen at their school. When we say *happen*, we mean in a big way. In 2018, Angela and Erin talked about how they had noticed in classrooms many Pinterest and Teachers Pay Teachers anchor charts that looked "cute" but on further examination were not very mathematical. Despite the previous professional learning sessions they provided on developing a mathematical mindset, they were still seeing evidence that what some of the teachers were doing was counter to that thinking. As a result, the mathematics coach requested funding for a cohort of teachers from the school to attend the two major mathematics education conferences that were to be held locally in Washington, D.C. The grant she wrote was funded, and as a result, 11 teachers from the school were able to attend the consecutive annual conferences of the National Council of Supervisors of Mathematics and the NCTM. They also attended a session by the authors of the book *The Whole School Agreement: Aligning Across and Within Grades to Build Student Success* (Karp et al., 2018). After the session, Angela and Erin met in the open meeting space of the convention center, and capitalizing on the enthusiasm of the 11 team members, they made a plan to enact the process of an MWSA. They agreed with the ideas shared in the session in thinking that having a similar common approach in mathematics instruction reduces the need for reteaching, provides familiarity for students, requires less cognitive demand when language and notation are not changing each year, and develops students' reasonableness and sense-making. It was after this debriefing that their journey really began.

After returning to their school, they jump-started the process by using the article "Establishing a Mathematics Whole-School Agreement" by Karp et al. (2016) and the two RTE articles for the elementary and middle grades (Karp et al., 2014, 2015). As a starting point, the entire staff—including specialized teachers (e.g., physical education, art, music, foreign language) as well as student support services team members (e.g., special educators, speech pathologists, school counselors, and occupational therapists)—considered the question "What mathematical knowledge do we want all young mathematicians at our school to leave with?" The immediate and decisive response was "We want *thinkers*—children who can *think through* hard problems, who persevere, and who can articulate their thinking."

The mathematics coach, Angela, embarked with grade-level collaborative learning teams (CLTs) on a yearlong set of discussions around their mathematics standards—the Virginia Mathematics Standards of Learning. To prepare for this work, the mathematics coach did a great deal of preliminary work setting up how they would progress through the state standards. They used a template similar to the one shared in this book in the Try It Out at the end of Chapter 1, which examines language, notation, representations, rules, generalizations, instructional strategies, and lesson structure. Each grade-level team, without judgment, took on the task of going through their practices, for each component of the template, and identified and recorded grade-level discrepancies. The time devoted to this work was approximately 20 minutes once a month during common mathematics planning time. Initially, there was substantial talk about how they were teaching and the language they used. They realized that there were myriad ways by which the teachers were referring to mathematical ideas, and they noticed that their language and other aspects of teaching were not as precise as they wanted and needed them to be to foster the deep student understanding they desired.

Figure 9.1 is an example of the language they generated as they started to unpack their second-grade state standards on key grade-level topics divided over their four instructional quarters. This visual is a record of the assortment of words—the nonjudgmental collection—of the language teachers in the same grade level noticed being used for the different Virginia state standards.

FIGURE 9.1 • LANGUAGE USED BY THE COLLABORATIVE LEARNING TEAM RELATED TO SECOND-GRADE STATE STANDARDS

	Quarter 1	Quarter 2	Quarter 3	Quarter 4
Language	Place Value Read, write, and identify place value; Name 10 more, 10 less, 100 more, 100 less; compare and order whole #s, rounding to the nearest ten • Decade numbers • Benchmarks • Anchor Numbers • Midpoint • Friendly Numbers • Sticks/Circles • Greater than /Less than vs. Bigger/Smaller • Reading inequalities like a sentence (both ways) Counting by 2s, 5s, 10s and backwards by 10s; even & odd • "Partner" (determining even and odd) • Skip counting is synonymous with counting by "x" • Multiples (Multiples of 2) Ordinal numbers • "Position" rather than "place" (not to confuse with place value)	Addition/Subtraction within 50 • Sum, addend, difference, subtrahend, minuend • Round • Nearest • Estimate - strategic choice of a number NOT a guess Symmetry • Line of symmetry • Same size and same shape (Congruent) • Reflection - mirror image 2D/3D figures • Vocabulary from State Standards (e.g., edges, vertices, faces) • Plane and solid figures (circles, spheres, squares, cubes, rectangles, rectangular prisms)	Fractions • Unit fraction • Equal parts of a whole • Fair share/equal share • A whole *divided* into equal parts • Halves, thirds, fourths, sixths, etc. • Names of models- set, region, or length	Time • Analog, digital clock • Clock face • Clockwise, Counter-clockwise Measurement • Length – nearest in. • Weight – nearest lb. Temperature • Celsius • Fahrenheit Data/Graphs • Pictograph • Bar Graph • Key • Scale Probability • Outcome • Prediction • Experiment

Source: Torpy and Russo (2019). Excerpt shared with permission.

Here is an example of CLT brainstorming of the collection of representations used at the fifth-grade level prior to the agreement process (see Figure 9.2).

Figure 9.3 shows the unveiling of each grade's work presented for review. With the support of the district's elementary mathematics specialist, Angela combed through the data to identify commonalities and discrepancies that needed to be resolved. In this way, the work was strategically shared from one grade to the entire faculty.

After all the grade-level groups had completed these "brain dumps" and noted the discrepancies, the full staff came back together to come to a consensus on areas of dissonance. For example, they identified that teachers were using nine different terms to identify the features of an open number line (e.g., anchors, benchmarks, friendly numbers, midpoints). This calibration work was carried out at a whole staff meeting in June 2019, where they looked at their curriculum and assessments and made decisions. Then, in August—the start of the school year with the 2019–2020 staff—the MWSA shown in Figure 9.4 was presented, which represents some example commitments made as part of the MWSA.

FIGURE 9.2 • REPRESENTATIONS USED BY THE COLLABORATIVE LEARNING TEAM RELATED TO THE FIFTH-GRADE STANDARDS

	Quarter 1	Quarter 2	Quarter 3	Quarter 4
Models & schema	Prime/Composite; Even/Odd: • Hundreds board/chart/grid • Blank 100's grid • Factors with a T-chart • T-chart for finding common multiples; list with commas for finding multiples Rounding decimals; equivalencies and Fraction/Decimal ordering • Open Number line • Blank 100's grids • Using fraction pieces for finding equivalent fractions (e.g., $\frac{8}{8} = 1$) Practical Problems (real-world situations) and Order of Operations • GEMS/GEMA • Tape diagrams • Not using multiple = signs in an equation	Decimal Computation (+/-): • base ten materials, decimal squares Fraction Computation in practical problems (+/-): • Area model • Length model Fraction Computation in practical problems (x): • Area model on grid paper • Number line	Decimal Computation (x/÷): • Area model • Partial quotients Area, Perimeter, Volume • Grid paper • Geoboard • One inch tiles • Containers/blocks Metric measurement • Chart • Meter sticks • Scales Elapsed time • Open number line	Transformations and subdividing polygons • Graph paper for transformations (vertical or horizontal) • Dynamic geometry software • tangrams • pattern blocks Fundamental Counting Principle • Table with sample space for 2 categories or options (e.g., pizza size - S, M, L; topping - P, C, V) Stem-and-Leaf and Line Plots • Blank template - count data points, make "fill in the blank spaces," write in data points Mean, median, mode, range • Use of a calculator • Mean as fair share (balance) using cubes Patterns • Tables • Template to predict nth term

Source: Torpy and Russo (2019). Excerpt shared with permission.

FIGURE 9.3 • THE MATHEMATICS WHOLE SCHOOL AGREEMENT–DISCOVERY SCHOOL STYLE

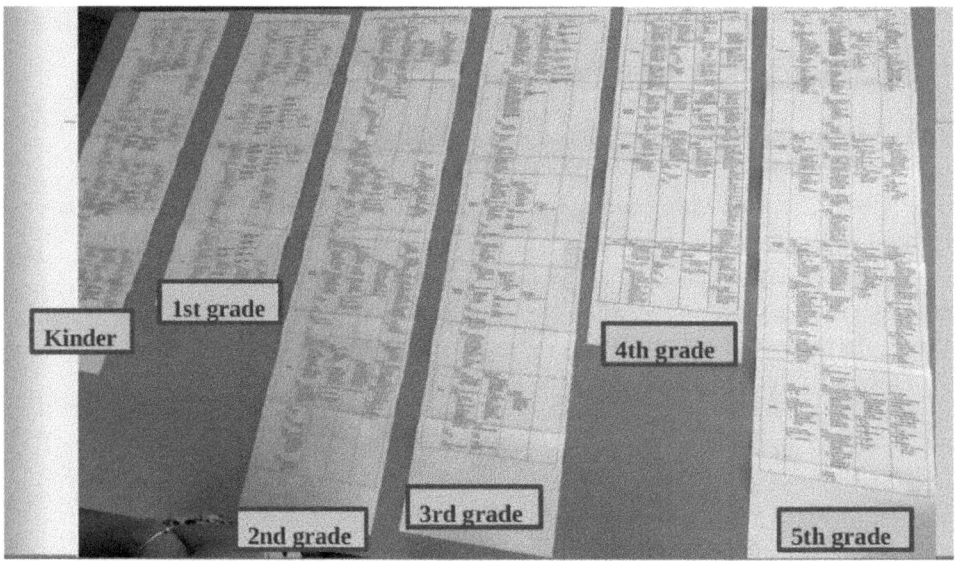

Source: Torpy and Russo (2019). Excerpt shared with permission.

FIGURE 9.4 • SHARED COMMITMENTS AT THE ELEMENTARY SCHOOL

Language We will...	Symbols & Notation We will...	Models & schema We will...	Rules We will...
• <u>name</u> properties - we will call them what they are • <u>name</u> parts of equations (to include subtraction) • differentiate between equation v expression • simplify fractions (NOT reduce) • refer to Base 10 as unit, rod, flat and cube when used concretely or pictorially • read whole numbers without saying "and" • read decimals with "and" and <u>not</u> "point" • read fractions without reading them as 3 out of 4 (Ss see 3 and 4 but not a single number) • use greater than and less than for comparing numbers NOT bigger and smaller • estimation = *strategic choice of a number!*	• emphasize fidelity to the "comma" to help mathematicians read numbers • not use the "alligator" for comparing numbers	• use a horizontal presentation of problems to invite multiple strategies • delay teaching algorithms to invite flexibility of thinking • present number lines as both vertical *and* horizontal • utilize an open number line as a strategy for rounding (midpoint, benchmarks)	• emphasize the use of benchmarks when comparing fractions • NOT use key words for problem solving • NOT use the "alligator" trick for comparing numbers

Source: Torpy and Russo (2019). Excerpt shared with permission.

Along the way, the team members realized that they were engaging in this process for a common cause—the students. The energy in the group was positive as they explored their current practice, how they needed to change, and how to practice that change. Because they were all part of the work, there was natural buy-in from all team members, and they felt empowered to collectively develop solutions. The grade-level teams then met in vertical teams at larger staff meetings toward the end of the year and found, at times, that there might be a person who bordered on being a naysayer. Again, positive peer pressure helped teachers who might have otherwise been stuck in their ways when the group would reflect and help those who felt resistant ask themselves "Why am I holding on to this?"

CORE MWSA IDEA

When we talk about language, we say, "mathematicians say it this way," thus framing the topic differently and engaging ourselves in empowering students as doers of mathematics.

To sustain continued implementation, the school refers to their MWSA frequently in their mathematics CLTs as they plan and launch new units. Teachers have gotten more comfortable holding one another accountable when non-MWSA language is used. They frame this with "Mathematicians say it this way," thus framing the topic in line with their MWSA.

The team decided that the next step in their MWSA team is to educate parents. They hope to have parents and guardians come to the school to observe their child during mathematics instruction—what they call a collaborative learning visit. They are optimistic that these visits give families the opportunity to ask questions of teachers, administrators, and coaches who act in the role of mathematics education docents. These visits are planned to conclude with family members indicating one actionable commitment or takeaway that they will carry out at home to support their student. This attention to the other educators in the children's lives is not only important but also powerful.

Empowered by their work and excited about the process, the team has now presented a recounting of this amazing story, called "A Journey Towards a Whole School Math Agreement," at a conference of the Virginia Council of Mathematics Specialists. Their goal was to empower the participants in the session to develop their own version by listening to the full accounting of the work required to reach an MWSA. Angela and Erin presented not only why they wanted to "agree" but also how others could learn from their approach. They described how to avoid stumbling blocks and how they now look ahead to hire the right new teachers to continue these efforts. They found that they needed to keep this process relevant and at the forefront of all decision-making and that just presenting a document with words as an agreement was not enough. They know this is a dynamic process that they must be living and breathing each day.

One Educator Spearheads an MWSA in His School

You must be the change you wish to see in the world.

—Mahatma Gandhi, social activist,
proponent of nonviolent action

An elementary instructional coach for a public school district in the Midwest took the lead initially on his own to create an MWSA. Reading had been an area of attention for the teachers in his school for about 10 years, with some basic attention given to mathematics. At times, teachers felt as if they were placing band-aids on issues or attempting to find quick fixes with mathematics. A much more thorough, intentional process was definitely called for. This seesawing of attention in the district between mathematics and reading felt very normal, with the weight often based on the perceived need identified by state standardized testing. This instructional coach heard about and loved the idea of an MWSA because it was about looking within, not relying on the state department of education or others to tell the school what it needed to do. Also, he felt that school success was about maximizing *all* academic areas—in the belief that they as educators rise and fall together. In this instructional coach's opinion, it was time to shift the focus to mathematics.

Their school team had previously developed graphic organizers to establish consistency in writing across the entire school, so he felt that they were primed to engage in the process of and enact an MWSA. There was a perfect storm of change, especially because they were also switching to new universal mathematics screeners, adopting new formative assessments, aligning to new pacing guides, and getting used to a new administration. Interestingly, the new principal had moved to the use of no prescribed mathematics curriculum. That change required teachers to make a huge cognitive shift. The instructional coach felt that the MWSA was the right approach at the right time to help support teachers and empower them.

Teachers at this school talked about being receptive, as they agreed that the discrepancies in instruction were an issue. There was evidence that some of the teachers were basing instruction on Teachers Pay Teachers resources and everyone was doing different things in their instruction. The Every Student Succeeds Act (2015) required the school to use resources based on "evidence-based activities, strategies, or interventions" (Kentucky Department of Education, 2020, n.p.), which was clearly not happening across the school. All trends in the school's academic and behavior data showed issues rooted in the need to refine mathematics instruction. For instance, they noticed that most of their behavior issues (i.e., students receiving office discipline referrals) were occurring during mathematics blocks. Also, the school's walk-through scores collected during mathematics by the instructional leader with the AdvancED/Cognia Effective Learning Environments Observation Tool® (eleot®; Davison, 2014) were consistently low in the categories of "supportive learning environment," "active learning environment," "progress monitoring and feedback," "high expectations environment," and, sadly, "equitable learning environment." The school had just started to use the Mathematics Classroom Observation Protocol for Practices (Gleason et al., 2017) in conjunction with the eleot to get a more rounded picture of the mathematics instructional culture. Collectively, by examining and analyzing these data, the group began to see that they had a crisis.

CORE MWSA IDEA

School success is about maximizing *all* academic areas—in the belief that they rise and fall together.

CORE MWSA IDEA

When school data reveal that most student discipline referrals occur during mathematics instruction, changes must be made!

Several faculty joined in to help lead the MWSA initiative as they recognized that they needed to "move from just doing what you want" to a cohesive approach through dynamic change. They started with grade-level teams that included related arts and special education teachers blended in. They all volunteered for representation on a mathematics committee to analyze the mathematics walk-through data using the eleot (Davison, 2014) and to unpack the state standards with a regional mathematics coordinator. They examined progressions, vertical alignment, the specific language of the standards, what mathematics was expected before and after each standard through Achieve the Core (2020) resources, and materials provided by a state mathematics cooperative. They identified some areas to work on, such as the design and delivery of instruction and building in active engagement and critical thinking. They decided to use a common mathematics discourse protocol so that students would talk about what mathematics they have learned.

The school's recent scores on the new benchmark test showed that the common formative assessments they had created in mathematics were not rigorous enough—that is, they didn't fully assess the intent and rigor of the standard and didn't always include components of the mathematical practices. Part of their ongoing MWSA is a critical analysis of the assessments they are creating horizontally in each grade level and how those connect vertically to what is expected before and after each grade level. The team now feels that the assessments they create are key to planning. They have initiated an assessment calendar/protocol for teachers, where they send their assessments to the elementary instructional coach leading the MWSA and he reviews them for alignment and clarity. He examines (a) if the assessment will provide meaningful information about students' learning, (b) if the results will provide meaningful opportunities for intervention, and (c) the general design of the instrument (no gotchas or tricks allowed!). The attention to mathematics instruction and assessment is helpful and is very much a part of how they are refining their systems and processes as part of the MWSA initiative.

This group of dedicated educators is also making the process their own in other ways. All teachers in the school have been trained in a discourse-heavy model of instruction. For them, it made sense to make discourse a uniting component of their MWSA. They asked, "How can we align discourse, mathematics talk, vocabulary, and language vertically? How can we encourage student engagement through giving them a chance to talk about mathematics?" A format they are considering can be seen in Figure 9.5.

FIGURE 9.5 • STRUCTURE FOR BUILDING STUDENT ENGAGEMENT THROUGH TALKING ABOUT MATHEMATICS

You: (Talk to yourself)
Start thinking about the problem on your own.
What's important?
When you imagine the situation, what is happening?
What strategies can you try?
What models, visuals, and drawings might help?
Y'all: (Talk to a partner)
Why did you use that approach?
How is your strategy like your partner's?
How is it different?
Explain your representations to your partner.
We: (Talk as a class)
What solution strategies did you use?
What do the materials, sketches, drawings, or symbols represent?
Reflection: (Talk with your partner)
What are the best strategies you heard in the class discussion?
Can you both try a different strategy?
What did you learn?

Source: Adapted from Green (2014).

While this group is still in the early stages of the MWSA process, they have embarked on the cognitive shift of sharing the responsibility for the mathematics success of each and every student in the school. This story portrays the power of a collective group of educators doing their due diligence and giving mathematics instruction the thoughtful attention it deserves!

A Mathematics Coach Takes the Lead: Helping Change Direction

If you do not change direction, you may end up where you are heading.

—Lao Tzu, philosopher

At Sigsbee Charter School in Key West, Florida, grades K–8, the mathematics coach led the development of an MWSA after a careful analysis of what was happening in the classrooms. The analysis included observations of every teacher who taught mathematics and discussions within each grade-level team where teachers shared issues they were struggling with in their classes. Additionally, the K–8 mathematics team, including the principal, analyzed their student achievement data and longitudinal trends and found that students were not making gains as they had hoped. In their analysis, they found that

- elementary mathematics lessons were modeled on the reading mini-lesson approach;
- the focus across grades was on skill development and not conceptual understanding or problem-solving;
- instruction was primarily teacher led, with little student engagement; and
- some mathematics content was inaccurate, and language was not precise.

The team agreed that something needed to be done to support students better. They followed the enactment described in Chapter 8 almost exactly. Since the time their MWSA was developed, the original mathematics coach had left the school, but a teacher who was part of the development process moved into the coaching position. We asked Megan Wise, a former third-grade teacher, to share her own thoughts and those of her colleagues about the MWSA approach and the long-term results, as she was part of the initial development and has since carried on with the process in her leadership role.

Teachers at the school found the MWSA to be helpful because all teachers now have a common language. When there are new teachers, there is a framework for helping them acclimate to the expectations for mathematics instruction in the school, such as not using the alligator mouth to teach inequalities, reading numbers correctly, and always writing fractions with the horizontal fraction bar.

Their MWSA helped the K–8 teachers write mathematics questions for warm-ups, using the reversibility, flexibility, and generalization Process Framework (see Chapter 6). Students see the same formats of tasks in every grade, so it is easier for teachers to facilitate instruction where students develop the ability to make generalizations. Students know what to expect, and they are ready for it!

Additionally, students would say things such as "We didn't learn that in grade ___." Teachers now are consistent in keeping their word charts (similar to a math word wall), and teachers in later grades often go to the previous grade teacher and borrow the word charts from that class. They show the charts to their current class, and the students recognize the words and the way they are written—it helps them connect their prior knowledge to what they are doing now.

The MWSA has influenced their evaluation systems in two ways. One, student assessments have a similar design across grades, and the reporting of the data from the assessments is consistent; families have found that to be helpful to better understand their own child's progress. Two, when the mathematics coach and the principal observe mathematics classes, the feedback from both is now consistent, rather than the principal and the coach focusing on different sets of criteria. Expectations are aligned.

As part of their MWSA, the mathematics team agreed on a lesson structure that focuses on promoting more student discourse. When teachers observe one another, they have an expectation of what they should see. This also places an accountability on each teacher to follow through with what was agreed on in their lessons. At the beginning of the year, when teachers are trying to establish the classroom culture and norms, there has been a learning curve for the students. Now, because of the relatively common lesson structure, students acclimate much faster and understand what their role is in the class and what their teacher's role is. Given that this school has a lot of student movement due to the large percentage of students from military families, who move in and out, the lesson consistency lets students who were in the school the previous year easily help the new students because they know what is going to happen—mathematics instruction is predictable.

Having the Standards for Mathematical Practice (SMP; Common Core State Standards Initiative, n.d.) as an agreed-on part of lesson structure has meant that students have more discussions in class and participate regularly in problem-solving tasks. The SMP as a significant focus area are a part of peer observations.

A side benefit occurred during the process of creating the MWSA. As Megan, the coach, said,

CORE MWSA IDEA

The MWSA builds an agreed-on focus on the standards for mathematical practice.

> Some of us didn't know that we didn't know! As we talked about language and generalizations, for example, we realized that we had some of our own misconceptions. It also promoted a reflection on your own practice so that you became more aware of things you might do without thinking—such as reading numbers wrong.

This provided an opportunity to explore mathematics content during the MWSA meetings.

With regard to professional development, the school has used the constructs of the MWSA to prioritize funding or decide on the type of professional learning experiences needed. Any professional development is aligned to the MWSA, so that teachers don't attend professional development that focuses on something different from the MWSA and teachers or administrators don't spend funds on things that will not support the MWSA. For example, there was a state conference for which the principal was willing to fund teacher attendance, but when the teachers looked at the ideas being presented in the conference sessions, they decided that it would not be a good use of their time. The focus of the conference was going to be on direct instruction, where the teacher does all of the talking and the instruction would be focused on skill development, not on developing meaning and conceptual understanding, both components of their MWSA.

The mathematics coach noted that it never felt difficult creating the MWSA, because everyone was working together. It helped to build a stronger team, one where teachers felt that they were focused on the same thing—helping students learn mathematics in a positive and supportive environment.

Having completed their MWSA, the mathematics coach offered one piece of advice. She felt that the MWSA should be revisited every year to determine what, if anything, needs to be revised. If your curriculum materials change, there may be some things that you want to reconsider. Or your state standards or assessments may change—anything can affect your MWSA.

Let's take a look at a classroom episode of two students at Sigsbee Charter School who were engaged in converting decimals to fractions.

> **Kaitlin:** I don't know how to write "four point two six" as a fraction.
>
> **Caleb:** You could do it if you said the number right.
>
> **Kaitlin:** What do you mean? That's how we always read the number in our class.
>
> **Caleb:** But when you read it like that, you can't tell what the number really is. You're just reading a bunch of numbers.
>
> **Kaitlin:** Okay, then how do you read it?
>
> **Caleb:** Easy! "Four and twenty-six hundredths."
>
> **Kaitlin:** Wow! That's so much easier! Now I know how to write it.

In this episode notice that the students continued their conversation without asking the teacher to "tell" them the answer. Having students work to come to a consensus can really engage them in authentic opportunities to construct viable arguments and critique the reasoning of others (as in SMP 3; NGA Center for Best Practices & CCSSO, 2010). Here the students are truly living the MWSA, as they are applying the use of appropriate and more precise mathematical language and questioning rules that do not seem quite right. We want our students to be empowered in these ways!

NEXT STEPS: YOU'RE READY FOR ACTION—CONSIDER CHANGE OFTEN

To improve is to change; to be perfect is to change often.

—Winston Churchill, former prime minister
of the United Kingdom

We say Bravo! to all the people who we reported on in this chapter as heroically engaging in this important work, as well as all others out there who are beginning an MWSA! But we said that this was the penultimate leg of the journey—which means that there is one last leg as we leave the process in your able hands. Your story is next. Revisit, revise, and revitalize.

Now that you've made your *math pact*, you are an official member of the MWSA Math Pack (#mathpact)!

REFERENCES

Achieve the Core. (n.d.). *Mathematics: Focus by grade level.* https://achievethecore.org/category/774/mathematics-focus-by-grade-level

Adams, T. L., Thangata, F., & King, C. (2005). Weigh to go! Exploring mathematical language. *Mathematics Teaching in the Middle School, 10*(9), 444–448.

Aguirre, J. M., Mayfield-Ingram, K., & Martin, D. B. (2013). *The impact of identity in K–8 mathematics learning and teaching: Rethinking equity-based practices.* National Council of Teachers of Mathematics.

Almarode, J., Fisher, D., Thunder, K., Hattie, J., & Frey, N. (2019). *Teaching mathematics in the visible learning classroom, grades K–2.* Corwin.

Almarode, J., Fisher, D., Thunder, K., Moore, S. D., Hattie, J., & Frey, N. (2019). *Teaching mathematics in the visible learning classroom, grades 3–5.* Corwin.

Andrews, D., & Kobett, B. M. (2017, July 18). *Connection to discourse: Word problems* [Paper presentation]. National Council of Teachers of Mathematics Discourse Institute. Baltimore, MD.

Arcavi, A. (2005). Developing and using symbol sense in mathematics. *For the Learning of Mathematics, 25*(5), 42–47.

Barlow, A. T., & Cates, J. M. (2007). The answer is 20 cookies, what is the question? *Teaching Children Mathematics, 13*(5), 252–255.

Barrera, F., & Santos, M. (2001). Student's use and understanding of different mathematical representation of tasks in problem-solving instruction. In R. Speiser, C. A. Maher, & C. N. Walter (Eds.), *Proceedings of the twenty-third annual meeting, North American Chapter of the International Group for the Psychology of Mathematics Education* (Vol. 1, pp. 459–466). Psychology of Mathematics Education.

Bay-Williams, J., & Kling, G. (2019). *Math fact fluency: 60+ games and assessment tools to support learning and retention.* Association for Supervision and Curriculum Development; National Council of Teachers of Mathematics.

Bay-Williams, J. M., & Livers, S. (2009). Supporting math vocabulary acquisition. *Teaching Children Mathematics, 16*(4), 238–246. https://doi.org/10.5951/teacchilmath.22.1.0020

Bay-Williams, J. M., & Martinie, S. L. (2015). Order of operations: The myth and the math. *Teaching Children Mathematics, 22*(1), 20–27.

Berry, R. Q., III. (2018). *Thinking about instructional routines in mathematics teaching and learning* (President's message). National Council of Teachers of Mathematics. https://www.nctm.org/News-and-Calendar/Messages-from-the-President/Archive/Robert-Q_-Berry-III/Thinking-about-Instructional-Routines-in-Mathematics-Teaching-and-Learning/

Berry, R. Q., III. (2019). *Examining equitable teaching using the mathematics teaching framework* (President's message). National Council of Teachers of Mathematics. https://www.nctm.org/News-and-Calendar/Messages-from-the-President/Archive/Robert-Q_-Berry-III/Examining-Equitable-Teaching-Using-the-Mathematics-Teaching-Framework/

Boaler, J. (2016). *Mathematical mindsets: Unleashing students' potential through creative math, inspiring messages and innovative teaching.* Jossey-Bass.

Boston, M. D., Candela, A. G., & Dixon, J. K. (2019). *Making sense of mathematics for teaching to inform instructional quality.* Solution Tree Press.

Brown, K. W., & Ryan, R. M. (2003). The benefits of being present: Mindfulness and its role in psychological well-being. *Journal of Personality and Social Psychology, 84*(4), 822–848. https://doi.org/10.1037/0022-3514.84.4.822

Bush, S. B., & Cook, K. L. (2019). *Step into STEAM, grades K–5: Your standards-based action plan for deepening mathematics and science learning.* Corwin; National Council of Teachers of Mathematics.

Cardone, T., & MTBoS. (2015). *Nix the tricks: A guide to avoiding shortcuts that cut out math concept development.* Creative Commons License.

Carpenter, T. P., Fennema, E., Franke, M. L., Levi, L., & Empson, S. B. (2014). *Children's mathematics: Cognitively guided instruction* (2nd ed.). Heinemann.

Carraher, D. W., Martinez, M. V., & Schliemann, A. D. (2008). Early algebra and mathematical generalization. *ZDM Mathematics Education, 40*, 3–22. https://doi.org/10.1007/s11858-007-0067-7

Cavey, L. O., & Kinzel, M. T. (2014). From whole number to invert and multiply. *Teaching Children Mathematics, 20*(6), 375–383. https://doi.org/10.5951/teacchilmath.20.6.0374

Center for Applied Special Technology. (2019). *About universal design for learning.* http://www.cast.org/our-work/about-udl.html#.Xchl5zNKg2w

Chirume, S. (2012). How does the use of mathematical symbols influence understanding of mathematical concepts by secondary school students? *International Journal of Social Sciences & Education, 3*(1), 35–46.

Chow, J., & Wehby, J. (2019). Effects of symbolic and nonsymbolic equal-sign intervention in second-grade classrooms. *The Elementary School Journal, 119*(4), 677–702. https://doi.org/10.1086/703086

Common Core State Standards Initiative. (n.d.). *Standards for mathematical practice.* http://www.corestandards.org/Math/Practice/

Common Core State Standards Initiative. (2016). *Frequently asked questions.* http://www.corestandards.org/about-the-standards/frequently-asked-questions/

Cramer, K., Monson, D., Whitney, S., Leavitt, S., & Wyberg, T. (2010). Dividing fractions and problem solving. *Mathematics Teaching in the Middle School, 15*(6), 338–346.

Cramer, K., Monson, D., Wyberg, T., Leavitt, S., & Whitney, S. (2010). *Models for initial decimal ideas. Teaching Children Mathematics, 16*(2), 106–117.

Crespo, S., Celedón-Pattichis, S., & Civil, M. (Eds.). (2017). *Access and equity: Promoting high-quality mathematics in pre-K–grade 2.* National Council of Teachers of Mathematics.

Crespo, S., Celedón-Pattichis, S., & Civil, M. (Eds.). (2018). *Access and equity: Promoting high-quality mathematics in grades 3–5.* National Council of Teachers of Mathematics.

Davison, M. (2014). *Effective learning environments observation tool (eleot).* AdvancED Source Research and Standards for Quality Schools.

Desmet, L., Grégoire, J., & Mussolin, C. (2010). Developmental changes in the comparison of decimal fractions. *Learning and Instruction, 20*(6), 521–532. https://doi.org/10.1016/j .learninstruc.2009.07.004

Dixon, J. (2019). Leading instruction by introducing academic vocabulary. *DNA Mathematics.* http://www.dnamath.com/blog-post/five-ways-we-undermine-efforts-to-increase-student-achievement-and-what-to-do-about-it-part-4-of-5/

Dixon, J. K. (2018, June 25). Just-in-time vs. just-in-case scaffolding: How to foster productive perseverance. *Houghton Mifflin Harcourt.* https://www.hmhco.com/blog/just-in-time-vs-just-in-case-scaffolding-how-to-foster-productive-perseverance

Dixon, J. K., Brooks, L. A., & Carli, M. R. (2019). *Making sense of mathematics for teaching: The small group.* Solution Tree Press.

Dougherty, B., Bryant, D. P., Bryant, B. R., Darrough, R. L., & Pfannenstiel, K. H. (2015). Developing concepts and generalizations to build algebraic thinking: The reversibility, flexibility, and generalization approach. *Intervention in School and Clinic, 50*(5), 273–281. https://doi .org/10.1177/1053451214560892

Dougherty, B., & Foegen, A. (2014). *Concept of variable: Screening and progress monitoring tool.* Iowa State University.

Dougherty, B. J. (2001). Access to algebra: A process approach. In H. Chick, K. Stacey, J. Vincent, & J. Vincent (Eds.), *The future of the teaching and learning of algebra* (pp. 207–213). University of Melbourne.

Dougherty, B. J. (2008). Measure up: A quantitative view of early algebra. In J. J. Kaput, D. W. Carraher, & M. L. Blanton. (Eds.), *Algebra in the early grades* (pp. 389–412). Lawrence Erlbaum. https://doi.org/10.4324/9781315097435-18

Dougherty, B. J., Bryant, D. P., Bryant, B. R., & Shin, M. (2016). Helping students with mathematics difficulties understand ratios and proportions. *TEACHING Exceptional Children, 49*(2), 96–105. https://doi.org/10.1177/0040059916674897

Dougherty, B. J., Bush, S. B., & Karp, K. S. (2017). Circumventing high school rules that expire. *Mathematics Teacher, 111*(2), 134–139. https://doi.org/10.5951/mathteacher.111.2.0134

Dougherty, B. J., DeLeeuw, W., & Foegen, A. (2017). *Algebra screening and progress monitoring project: Concept of variable.* Iowa State University.

Dougherty, B. J., & Foegen, A. (2011). *Evaluation of the mathematics project for special education and general education teachers* (Report prepared for the Iowa Department of Education). Iowa State University.

Dougherty, B. J., Foegen, A., & DeLeeuw, W. (2017). *Concept of variable: Screening and progress monitoring tool, 2016 version.* Iowa State University.

Ellis, A. B. (2011). Generalizing-promoting actions: How classroom collaborations can support students' mathematical generalizations. *Journal for Research in Mathematics Education, 42*(4), 308–345. https://doi.org/10.5951/jresematheduc.42.4.0308

Every Student Succeeds Act, 20 U.S.C. § 6301 (2015). www.congress.gov/114/plaws/publ95/PLAW-114publ95.pdf

Falkner, K. P., Levi, L., & Carpenter, T. P. (1999). Children's understanding of equality: A foundation for algebra. *Teaching Children Mathematics, 6*(4), 232–236.

Fennell, F. M., Kobett, B. M., & Wray, J. A. (2017). *The formative 5: Everyday assessment techniques for every math classroom.* Corwin; National Council of Teachers of Mathematics.

Fey, J. (1990). Quantity. In L. A. Steen (Ed.), *On the shoulders of giants: New approaches to numeracy* (pp. 61–94). National Academies Press.

Gleason, J., Livers, S., & Zelkowski, J. (2017). Mathematics classroom observation protocol for practices (MCOP2): A validation study. *Investigations in Mathematics Learning, 9*(3), 111–129. https://doi.org/10.1080/19477503.2017.1308697

Gondo, M., Patterson, K. D., & Trujillo Palacios, S. (2013). Mindfulness and the development of a readiness for change. *Journal of Change Management, 13*(1), 36–51. https://doi.org/10.1080/14697017.2013.768431

Green, E. (2014, July 27). Why do Americans stink at math? *The New York Times,* p. 23. https://www.nytimes.com/2014/07/27/magazine/why-do-americans-stink-at-math.html

Gregg, J., & Gregg, D. U. (2007). Measurement and fair-sharing models for dividing fractions. *Mathematics Teaching in the Middle School, 12*(9), 490–496.

Hardiman, M. M. (2011). *Brain-targeted teaching.* https://braintargetedteaching.org/climate.cfm

Hardiman, M. M. (2012). *The brain-targeted teaching model for 21st century schools.* Corwin.

Hattie, J. (2018). Hattie ranking: 252 influences and effect sizes related to student achievement. *Visible Learning.* https://visible-learning.org/hattie-ranking-influences-effect-sizes-learning-achievement/. https://doi.org/10.4324/9780429485480

Hattie, J., Fisher, D., Frey, N., Gojak, L. M., Moore, S. D., & Mellman, W. (2016). *Visible learning for mathematics, grades K–12: What works best to optimize student learning.* Corwin.

Huinker, D., & Bill, V. (2017). *Taking action: Implementing effective mathematics teaching practices in K–grade 5.* National Council of Teachers of Mathematics.

Ivy, J., Bush, S. B., & Dougherty, B. J. (2020). Stacking the deck: Reversibility and reasoning. *Mathematics Teacher: Learning and Teaching PK–12, 113*(1), 65–68. https://doi.org/10.5951/MTLT.2019.0027

Iyengar, S. (2011). *TED talk.* https://www.ted.com/talks/sheena_iyengar_choosing_what_to_choose/transcript?language=en#t-225497

Iyengar, S., & Lepper, M. (2000). When choice is demotivating: Can one desire too much of a good thing? *Journal of Personality and Social Psychology, 79*(6), 995–1006. https://doi.org/10.1037/0022-3514.79.6.995

Jansen, A. (2020). *Rough draft math: Revising to learn.* Stenhouse.

Karp, K. S., Bush, S. B., & Dougherty, B. J. (2014). 13 rules that expire. *Teaching Children Mathematics, 21*(1), 18–25. https://doi.org/10.5951/teacchilmath.21.1.0018

Karp, K. S., Bush, S. B., & Dougherty, B. J. (2015). 12 math rules that expire in the middle grades. *Mathematics Teaching in the Middle School, 21*(4), 208–215. https://doi.org/10.5951/mathteacmiddscho.21.4.0208

Karp, K. S., Bush, S. B., & Dougherty, B. J. (2016). Establishing a mathematics whole-school agreement. *Teaching Children Mathematics, 23*(2), 61–63. https://doi.org/10.5951/teacchilmath.23.2.0061

Karp, K. S., Bush, S. B., & Dougherty, B. J. (2018, April). *The whole school agreement: Aligning across and within grades to build student success* [Paper presentation]. National Council of Supervisors of Mathematics annual meeting, , Washington, DC, United States.

Karp, K. S., Bush, S. B., & Dougherty, B. J. (2019). Avoiding the ineffective keyword strategy. *Teaching Children Mathematics*, 25(7), 428–435. https://doi.org/10.5951/teacchilmath.25.7.0428

Kentucky Department of Education. (2020). *Evidence-based practices.* https://education.ky.gov/school/evidence/Pages/default.aspx

Kieran, C. (2007). Learning and teaching algebra at the middle school through college levels: Building meaning for symbols and their manipulation. In F. K. Lester Jr., (Ed.), *Second handbook of research on mathematics teaching and learning* (pp. 707–762). Information Age. https://doi.org/10.1163/9789087901127_003

Kobett, B., & Karp, K. (2020). *Strengths-based teaching and learning in mathematics: 5 teaching turnarounds for grades K–6.* Corwin; National Council of Teachers of Mathematics.

Kobett, B. M., Miles, R. H., & Williams, L. A. (2018). *The mathematics lesson-planning handbook, grades K–2: Your blueprint for building cohesive lessons.* Corwin; National Council of Teachers of Mathematics.

Kreisberg, H. (2018, April 8). The era of resource abundance and what we can do about it. *QED.* https://medium.com/q-e-d/the-era-of-resource-abundance-and-what-we-can-do-about-it-89e8b953573e

Kreisberg, H. (2019, April 7). The era of resource abundance part II: How to navigate through the crap to find the rich and useful tasks. *QED.* https://medium.com/q-e-d/the-era-of-resource-abundance-part-ii-how-to-navigate-through-the-crap-to-find-the-rich-and-d7a9659f4e07

Küchemann, D. (1978). Children's understanding of numerical variables. *Mathematics in School*, 7(4), 23–26.

Kurz, T. L. (2013). Using technology to balance algebraic explorations. *Teaching Children Mathematics*, 19(9), 554–562. https://doi.org/10.5951/teacchilmath.19.9.0554

Lannin, J., Ellis, A. B., & Elliott, R. (2011). *Developing essential understanding of mathematical reasoning for teaching mathematics in prekindergarten–grade 8.* National Council of Teachers of Mathematics.

Larson, M. (2017, July 14). *What constitutes an effective collaborative team?* (President's message). National Council of Teachers of Mathematics. https://www.nctm.org/News-and-Calendar/Messages-from-the-President/Archive/Matt-Larson/What-Constitutes-an-Effective-Collaborative-Team_/

Laski, E. V., Jor'dan, J. R., Daoust, C., & Murray, A. K. (2015). What makes mathematics manipulatives effective? Lessons from cognitive science and Montessori education. *SAGE Open*, 5(2), 1–8. https://doi.org/10.1177/2158244015589588

Leavy, A., Hourigan, M., & McMahon, A. (2013). Early understanding of equality. *Teaching Children Mathematics*, 20(4), 247–252. https://doi.org/10.5951/teacchilmath.20.4.0246

Leinwand, S. (1994). Four teacher-friendly postulates for thriving in a sea of change. *Mathematics Teacher*, 87, 392–393.

Lesh, R., Post, T., & Behr, M. (1987). Representations and translations among representations in mathematics learning and problem solving. In C. Janvier (Ed.), *Problems of representation in the teaching and learning of mathematics* (pp. 33–40). Lawrence Erlbaum.

Lesseig, K. (2016). Conjecturing, generalizing and justifying: Building theory around teacher knowledge of proving. *International Journal for Mathematics Teaching and Learning, 17*(3), 1–31.

Lewin, K. (1947). Group decisions and social change. In T. M. Newcomb & E. L. Hartley (Eds.), *Readings in social psychology* (pp. 197–211). Holt, Rinehart & Winston.

Li, Y. (2008). What do students need to learn about division of fractions? *Mathematics Teaching in the Middle School, 13*(9), 546–552.

Linchevski, L., & Livneh, D. (1999). Structure sense: The relationship between algebraic and numerical contexts. *Educational Studies in Mathematics, 40*(2), 173–196. https://doi .org/10.1023/A:1003606308064

Livers, S., & Bay-Williams, J. M. (2014). Vocabulary support: Constructing (not obstructing) meaning). *Mathematics Teaching in the Middle School, 20*(3), 152–159. https://doi.org/10.5951/ mathteacmiddscho.20.3.0152

Livers, S. D., Harbour, K. E., & Fowler, L. (2019). Danger! Animals in the mathematics classroom. *Teaching Children Mathematics, 25*(7), 406–414. https://doi.org/10.5951/teacchilmath.25.7.0406

Luscombe, B. (2019, September 16). Former U.N. ambassador Samantha Power has a lot of stories to tell. *Time, 194*(10), 22–23. https://time.com/5669511/samantha-power-interview/

Maccini, P., & Ruhl, K. L. (2000). Effects of a graduated instructional sequence on the algebraic subtraction of integers by secondary students with learning disabilities. *Education & Treatment of Children, 23*(4), 465–489.

Mann, R. L. (2004). Balancing act: The truth behind the equals sign. *Teaching Children Mathematics, 11*(2), 65–69.

Martinie, S. (2014). Decimal fractions: An important point. *Mathematics Teaching in the Middle School, 19*(7), 420–429. https://doi.org/10.5951/mathteacmiddscho.19.7.0420

Matthews, P. G., & Fuchs, L. S. (2018). Keys to the gate? Equal sign knowledge at second grade predicts fourth-grade algebra competence. *Child Development, 91*(1), 1–15. https://doi.org/10.1111/ cdev.13144

Matthews, P. G., & Rittle-Johnson, B. (2009). In pursuit of knowledge: Comparing self-explanations, concepts, and procedures as pedagogical tools. *Journal of Experimental Child Psychology, 104*(1), 1–21. https://doi.org/10.1016/j.jecp.2008.08.004

McCaffrey, T. (2016, June 6). *Rethinking the gradual release of responsibility model.* https://www .nctm.org/Publications/Mathematics-Teaching-in-Middle-School/Blog/Rethinking-the-Gradual-Release-of-Responsibility-Model/

McNeil, N. M., Uttal, D. H., Jarvin, L., & Sternberg, R. J. (2009). Should you show me the money? Concrete objects both hurt and help performance on mathematics problems. *Learning and Instruction, 19*(2), 171–184. https://doi.org/10.1016/j.learninstruc.2008.03.005

Miles, R. H., Kobett, B. M., & Williams, L. A. (2018). *The mathematics lesson-planning handbook, grades 3–5: Your blueprint for building cohesive lessons.* Corwin; National Council of Teachers of Mathematics.

Miller, S. P., Mercer, C. D., & Dillon, A. S. (1992). CSA: Acquiring and retaining math skills. *Intervention in School and Clinic, 28*(2), 105–110. https://doi.org/10.1177/105345129202800206

Miller Bennett, V. (2017). Understanding the meaning of the equal sign. *New England Journal of Mathematics, 50*(1), 18–25.

Molina, M., & Ambrose, R. C. (2006). Fostering relational thinking while negotiating the meaning of the equals sign. *Teaching Children Mathematics, 13*(2), 111–117.

Moreno, R., Ozogul, G., & Reisslein, M. (2011). Teaching with concrete and abstract visual representations: Effects on students' problem solving, problem representations, and learning perceptions. *Journal of Educational Psychology, 103*(1), 32–47. https://doi.org/10.1037/a0021995

Moschkovich, J. (2019). Codeswitching and mathematics learners: How hybrid language practices provide resources for student participation in mathematical practices. In J. MacSwan & C. J. Faltis (Eds.), *Critical perspectives on codeswitching in classroom settings: Language practices for multilingual teaching and learning* (pp. 88–112). Routledge. https://doi.org/10.4324/9781315401102-4

National Council of Teachers of Mathematics. (2000). *Principles and standards for school mathematics.* https://www.nctm.org/uploadedFiles/Standards_and_Positions/PSSM_ExecutiveSummary.pdf

National Council of Teachers of Mathematics. (2010–2013). *Developing essential understanding series.*

National Council of Teachers of Mathematics. (2013–2019). *Putting essential understanding into practice series.*

National Council of Teachers of Mathematics. (2014a). *Principles to actions: Ensuring mathematical success for all.*

National Council of Teachers of Mathematics. (2014b). *Procedural fluency in mathematics: A position of the National Council of Teachers of Mathematics.* https://www.nctm.org/Standards-and-Positions/Position-Statements/Procedural-Fluency-in-Mathematics/

National Council of Teachers of Mathematics. (2020). *Catalyzing change in early childhood and elementary mathematics: Initiating critical conversations.*

National Governors Association Center for Best Practices, & Council of Chief State School Officers. (2010). *Common core state standards for mathematics.* http://www.corestandards.org/Math/

Okazaki, C., Zenigami, F., & Dougherty, B. J. (2006). Measure up: A different view of elementary mathematics. In S. Smith & S. Smith (Eds.), *Teachers engaged in research* (pp. 135–152). Information Age.

Opfer, V. D., Kaufman, J. H., & Thompson, L. E. (2016). *Implementation of K–12 state standards for mathematics and English language arts and literacy: Findings from the American Teacher Panel.* RAND Corporation.

Pape, S. J., & Tchoshanov, M. A. (2001). The role of representation(s) in developing mathematical understanding. *Theory Into Practice, 40*(2), 118–127. https://doi.org/10.1207/s15430421tip4002_6

Parrish, S. (2014). *Number talks: Whole number computation, grades K–5.* Math Solutions.

Parrish, S., & Dominick, A. (2016). *Number talks: Fractions, decimals and percentages.* Math Solutions.

Peltier, C., & Vannest, K. J. (2018). Using the concrete representational abstract (CRA) instructional framework for mathematics with students with emotional and behavioral disorders. *Preventing School Failure: Alternative Education for Children and Youth, 62*(2), 73–82. https://doi.org/10.1080/1045988X.2017.1354809

Perlwitz, M. D. (2004). Two students' constructed strategies to divide fractions. *Mathematics Teaching in the Middle School, 10*(3), 122–126.

Philipp, R. A., Cabral, C., & Schappelle, B. P. (2005). *IMAP CDROM: Integrating mathematics and pedagogy to illustrate children's reasoning* [Computer software]. Pearson.

Polikoff, M., & Dean, J. (2019). *The supplemental curriculum bazaar: Is what's online any good?* Thomas B. Fordham Institute. https://fordhaminstitute.org/national/research/supplemental-curriculum-bazaar

Rambia, S. (2002). A new approach to an old order. *Mathematics Teaching in the Middle School, 8*(4), 193–195.

Reinhart, S. C. (2000). Never say anything a kid can say! *Mathematics Teaching in the Middle School, 5*(8), 478–483.

Richardson, K. (1998–2015). *Developing number concepts* [Book series]. Math Perspectives Teacher Development Center.

Richardson, K. (2003). *Assessing math concepts* [Book series]. Math Perspectives Teacher Development Center.

Rittle-Johnson, B., Matthews, P. G., Taylor, R. S., & McEldoon, K. L. (2011). Assessing knowledge of mathematical equivalence: A construct-modeling approach. *Journal of Educational Psychology, 103*(1), 85–104. https://doi.org/10.1037/a0021334

Rosenbaum, D., More, E., & Steane, P. (2018). Planned organizational change management: Forward to the past? An exploratory literature review. *Journal of Organizational Change Management, 31*(2), 286–303. https://doi.org/10.1108/JOCM-06-2015-0089

Rubenstein, R. N., & Thompson, D. R. (2002). Understanding and supporting children's mathematical vocabulary development. *Teaching Children Mathematics, 9*(2), 107–112.

SanGiovanni, J. J. (2016a). *Mine the gap for mathematical understanding, grades K–2: Common holes and misconceptions and what to do about them.* Corwin.

SanGiovanni, J. J. (2016b). *Mine the gap for mathematical understanding, grades 3–5: Common holes and misconceptions and what to do about them.* Corwin.

SanGiovanni, J. J., Katt, S., & Dykema, K. (2020). *Productive math struggle: A six point action plan for fostering perseverance.* Corwin.

Santelises, S. B., & Dabrowski, J. (2015). *Checking in: Do classroom assignments reflect today's higher standards?* Education Trust.

Shaughnessy, M. M. (2011). Identify fractions and decimals on a number line. *Teaching Children Mathematics, 17*(7), 428–434.

Skemp, R. (1978). Relational understanding and instrumental understanding. *Arithmetic Teacher, 26*(3), 9–15.

Smith, M. S., Bill, V., & Sherin, M. G. (2019). *The 5 practices in practice: Successfully orchestrating mathematical discussion in your elementary school classroom.* Corwin.

Smith, M. S., Steele, M. D., & Raith, M. L. (2017). *Taking action: Implementing effective mathematics teaching practices.* National Council of Teachers of Mathematics.

Smith, M. S., & Stein, M. K. (1998). Selecting and creating mathematical tasks: From research to practice. *Mathematics Teaching in the Middle School, 3*(5), 344–350.

Smith, M. S., & Stein, M. K. (2011). *5 practices for orchestrating productive mathematics discussions*. Corwin; National Council of Teachers of Mathematics.

Smith, M. S., & Stein, M. K. (2018). *5 Practices for orchestrating productive mathematics discussions* (2nd ed.). Corwin; National Council of Teachers of Mathematics.

Squire, L. (2004). Memory systems of the brain: A brief history and current perspective. *Neurobiology of Learning and Memory, 82*(3), 171–177. https://doi.org/10.1016/j.nlm.2004.06.005

Staples, M. E., Bartlo, J., & Thanheiser, E. (2012). Justification as a teaching and learning practice: Its (potential) multifaceted role in middle grades mathematics classrooms. *Journal of Mathematical Behavior, 31*(4), 447–462. https://doi.org/10.1016/j.jmathb.2012.07.001

Stephens, A. C., Knuth, E. J., Blanton, M. L., Isler, I., Gardiner, A. M., & Marum, T. (2013). Equation structure and the meaning of the equal sign: The impact of task selection in eliciting elementary students' understandings. *Journal of Mathematical Behavior, 32*(2), 173–182. https://doi.org/10.1016/j.jmathb.2013.02.001

Torpy, A. & Russo, E. (2019, October 4). *Our story of discovery: A journey towards a whole school math agreement* [Paper presentation]. Virginia Council of Mathematics Specialists 8th Annual Conference. Fredericksburg, VA, United States.

Usiskin, Z. (2015). What does it mean to understand some mathematics? In J. C. Sung (Ed.), *Selected regular lectures from the 12th International Congress on Mathematical Education* (pp. 821–841). Springer. https://doi.org/10.1007/978–3-319-17187-6_46

Van de Walle, J. A., Karp, K. S., & Bay-Williams, J. M. (2019). *Elementary and middle school mathematics: Teaching developmentally* (10th ed.). Pearson.

Van de Walle, J. A., Karp, K. S., Lovin, L. H., & Bay-Williams, J. M. (2018). *Teaching student centered mathematics: Teaching developmentally in grades 3–5*. Pearson.

Van de Walle, J. A., Lovin, L. H., Karp, K. S., & Bay-Williams, J. M. (2018). *Teaching student-centered mathematics: Teaching developmentally in grades preK–2*. Pearson.

Weir, A. (2014). *The Martian*. Crown.

Wiseman, L., (2017). *Multipliers, revised and updates: How the best leaders make everyone smarter*. Harper Business.

Wiseman, L., Allen, L., & Foster, E. (2013). *The multiplier effect: Tapping the genius inside our schools*. Corwin.

Yeh, C., Ellis, M., & Hurtado, C. K. (2017). *Reimagining the mathematics classroom: Creating and sustaining productive learning environments*. National Council of Teachers of Mathematics.

Zager, T. (2017). *Becoming the math teacher you wish you'd had: Ideas and strategies from vibrant classrooms*. Stenhouse.

INDEX

ALL students should have the opportunity to be successful in mathematics!

Trusted experts in mathematics education offer clear and practical guidance to help students move from surface to deep mathematical understanding, from procedural to conceptual learning, and from rote memorization to true comprehension. Through books, videos, consulting, and online tools, we offer a truly blended learning experience that helps you demystify mathematics for students.

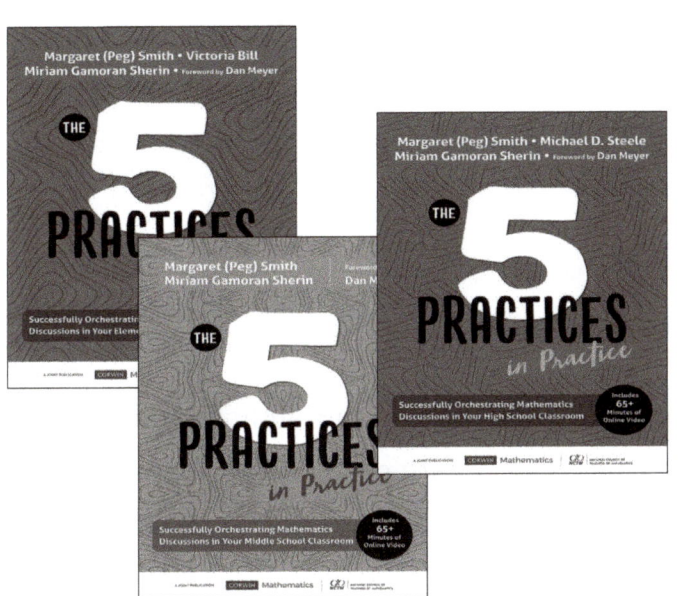

MARGARET (PEG) SMITH, VICTORIA BILL, MIRIAM GAMORAN SHERIN, MICHAEL D. STEELE

Enhance your fluency in the five practices to bring powerful discussions of mathematical concepts to life in your classroom.

Elementary, Middle School, High School

JOHN J. SANGIOVANNI, SUSIE KATT, KEVIN J. DYKEMA

Provides an essential plan for embracing productive perseverance in mathematics by guiding teachers through six specific actions—including valuing, fostering, building, planning, supporting, and reflecting on struggle.
Grades K-12

PETER LILJEDAHL

Help teachers implement 14 optimal practices for thinking that create an ideal setting for deep mathematics learning to occur.
Grades K-12

CORWIN **Mathematics**

A SAGE Publishing Company

Helping educators make the greatest impact

CORWIN HAS ONE MISSION: to enhance education through intentional professional learning.

We build long-term relationships with our authors, educators, clients, and associations who partner with us to develop and continuously improve the best evidence-based practices that establish and support lifelong learning.

NATIONAL COUNCIL OF
TEACHERS OF MATHEMATICS

The National Council of Teachers of Mathematics supports and advocates for the highest-quality mathematics teaching and learning for each and every student.